TECHNICAL
REPORT

Understanding and Reducing Off-Duty Vehicle Crashes Among Military Personnel

Liisa Ecola, Rebecca L. Collins, Elisa Eiseman

Prepared for the Defense Safety Oversight Council

NATIONAL DEFENSE RESEARCH INSTITUTE

The research described in this report was prepared for the Defense Safety Oversight Council. The research was conducted in the RAND National Defense Research Institute, a federally funded research and development center sponsored by the Office of the Secretary of Defense, the Joint Staff, the Unified Combatant Commands, the Navy, the Marine Corps, the defense agencies, and the defense Intelligence Community under Contract W74V8H-06-C-0002.

Library of Congress Control Number: 2010936945

ISBN: 978-0-8330-5021-2

Published 2010 by the RAND Corporation
1776 Main Street, P.O. Box 2138, Santa Monica, CA 90407-2138
1200 South Hayes Street, Arlington, VA 22202-5050
4570 Fifth Avenue, Suite 600, Pittsburgh, PA 15213-2665
RAND URL: http://www.rand.org/
To order RAND documents or to obtain additional information, contact
Distribution Services: Telephone: (310) 451-7002;
Fax: (310) 451-6915; Email: order@rand.org

Preface

The Department of Defense (DoD) established the Defense Safety Oversight Council (DSOC) in July 2003. DSOC was chartered to review accident and incident trends, ongoing safety initiatives, and private-sector and other governmental agency best practices and to make recommendations to the Secretary of Defense for safety improvement policies, programs, and investments with a goal of reducing military accidents and the attendant loss of life, workdays, and equipment. DSOC seeks to better track the causal factors and costs of accidents through the use of metrics and tools established specifically for that purpose and to execute task force–nominated initiatives directed at reducing the number and severity of preventable accidents by 75 percent across DoD within five years.

The Private Motor Vehicle (PMV) Task Force (TF) is one of nine flag/general officer level, joint task forces charged with achieving the goals of DSOC. It focuses on reducing the number and severity of preventable, privately owned motor vehicle mishaps among U.S. service members. PMV TF is composed of operation leaders, traffic safety officers, and certified safety professionals drawn from across the services and military departments. The task force is charged with reducing preventable PMV-related mishaps through the development and integration of DoD-wide mishap mitigation strategies; by evaluating and making recommended changes to service traffic policy and safety regulations; by reviewing and making recommendations to service training programs in the areas of driver improvement, remediation, and motorcycle operation; and by overseeing the development and implementation of DoD-level safety campaigns and enforcement strategies.

This report supports the work of PMV TF. It provides an evidence base so that the task force can target these efforts most effectively. The report identifies and reviews evidence-based literature on driving behaviors (fatigue, distracted driving, combat driving, and other risky behaviors) and cultural influences on driving behaviors. It focuses on such influences particularly with respect to young drivers, service members, and motorcycle riders.

This report makes suggestions based on the literature regarding how best to reduce crash fatalities within the military. Because PMV TF focuses on "preventable" crashes, the report addresses what is known about areas over which the task force can exert some control: the demographic groups most likely to be in crashes, driver behaviors that are most dangerous and how they affect the occurrence and severity of crashes, personality traits associated with risk-taking, and the effectiveness of measures such as safety campaigns, law enforcement, and driver training. The team did not examine other factors that contribute to vehicle crashes, such as weather, road conditions, and vehicle performance and safety features, with the exception of reviewing some issues specific to motorcycling such as engine size. Certain issues that were mentioned in the scope of the work, such as the impact of combat-driving behavior, were little discussed in the existing empirical literature, and the team did not undertake new research in these areas.

These suggestions are based on the team's review of the findings in the evidence-based literature and not on the work of PMV TF itself (that is, this report did not undertake any rigorous evaluation of the many initiatives put forward by the task force). We hope that these can be integrated into the work of the task force as it works to reduce vehicle crashes and their severity. This report should also be of interest to the broader traffic safety community outside the task force. Comments are welcome and may be addressed to Martin_Wachs@rand.org.

This research was sponsored by DSOC through the National Defense Center for Energy and Environment (NDCEE) and conducted within the Forces and Resources Policy Center of the RAND National Defense Research Institute, a federally funded research and development center sponsored by the Office of the Secretary of Defense, the Joint Staff, the Unified Combatant Commands, the Navy, the Marine Corps, the defense agencies, and the defense Intelligence Community.

For more information on the RAND Forces and Resources Policy Center, see http://www.rand.org/nsrd/about/frp.html or contact the Director (contact information is provided on the web page).

Contents

Figures

Tables

Summary

This report reviews a wide variety of evidence regarding traffic safety in the United States, with specific reference to military personnel. The report has two broad organizing themes: first, who is most at risk for being in a vehicle crash; and second, what measures can be taken to alleviate this risk. In terms of such measures, the report focuses on safety interventions and attempts to change driver behavior and decisions.

General Trends in Vehicle Crash Fatalities

Overall, driving has become safer over the past 20 years. Three common measures—total fatalities, fatalities per 100,000 population, and fatalities per 100 million miles driven—all show substantial decreases since the early 1990s. A variety of factors seem to have contributed to this decline: better vehicle safety features, better road safety features, decreases in teenage drunk driving, more seat belt use, and at least in the past several years, fewer vehicle miles traveled. These impressive trends mean that fewer Americans are killed or injured in vehicle crashes and that, mile for mile, the chance of being in a serious crash are fairly low (between one and two people are killed for every 100 million miles driven).

In contrast, motorcycle riding, a topic of particular interest to the military, is becoming more dangerous. Total motorcycle fatalities and fatalities per mile driven are much higher now than in the mid-1990s. The causes of this trend are less certain, but several factors seem to contribute: increases in motorcycle use; the rising popularity of sport bikes and supersports, which combine powerful engines with lightweight frames; and the trend away from universal to partial helmet laws (under which only riders less than a certain age are required to wear helmets). Fatalities are rising most significantly among men over 40, but all ages and bike types have seen increases.

For military personnel, overall crash fatality rates are generally higher than for the U.S. population, but similar to those of men of comparable ages. Young men are at the highest risk for death in vehicle crashes of any demographic group, largely because of their propensity to take more risks behind the wheel. Indeed, while crash death rates for young women are highest when they are teenagers, for men the risks continue to rise into their early 20s. The main difference between the military and civilian population is the proportion of military crash fatalities on motorcycles—the U.S. rate is currently about 15 percent of fatalities, while in some military branches the rate is on average 35–40 percent.

Drivers at Risk

A number of driver behaviors contribute to both the occurrence and severity of crashes. Both drunk driving and speeding are responsible for about one-third of fatal crashes (crashes may be due to more than one factor), while about half of fatal crashes involve vehicle occupants who were not wearing seat belts. Distracted driving (for example, talking on a cell phone or texting) and driving while fatigued seem to contribute to fewer crashes, but data to study these factors are more limited, so estimates vary about what proportion of crashes they cause.

Young men are more at risk than older men and than women, in part because they take more risks, and in part because young drivers are more vulnerable to certain risks than are more-experienced drivers. Evidence shows that young people are more susceptible to the effects of alcohol, and they are more likely to be in crashes at lower blood alcohol concentration (BAC) levels than are older drivers. Young men are disproportionately likely to be in crashes because of speeding, and they are less likely to wear seat belts.

It is less clear whether young people are also more vulnerable to the effects of distracted driving—while some studies indicate they are, others have cautioned that these studies do not account for the fact that young people are more likely to use cell phones than are older drivers. The research related to distracted driving suggests that the main danger of both using a cell phone and texting is slower reaction time, which increases stopping distance. The research is also clear that hand-held and hands-free phones present equal dangers.

Military populations have many of the same risk factors as do civilian populations: Young people are more likely to be in crashes than are people over 35, motorcycle riders are at higher risk than car drivers, and those who drink heavily and do not regularly wear seat belts are at greater risk. Detailed studies of military vehicle crashes and deaths have found other risk factors as well, although the effects tend to be smaller: People with only a high school education or less are at greater risk than those with some college, unmarried people are at higher risk than married ones, certain military occupations seem to be at higher risk (although these vary among studies), and those who have deployed have a higher risk than those who have not.

The deployment risk, which is corroborated by multiple studies from Vietnam and the first Gulf War and one study of Operation Enduring Freedom/Operation Iraqi Freedom, is not well understood. It is possible that military personnel who deploy are by nature more prone to take risks than are those who do not, meaning that the deployment itself is not the cause of the increased crash risk. The risk has also been attributed to posttraumatic stress disorder, greater risk-taking, heavier drinking, and the possibility that deployment injuries may make it harder to survive a crash. There is not enough information to say definitively whether some of these deaths might be caused by fatigue or attributable to suicide.

Although evidence comes from only one study of enlisted Marines, the profile of who in the military is at greatest risk of being in a car crash seems quite different from who might be in a motorcycle crash. Those in car crashes tended to be fairly young and single, but with little difference by race, gender, or occupation. Those at highest risk for motorcycle crashes were white men, probably in their mid-20s to early 30s, and clustered in certain occupations. Having entered the Marines with a felony waiver was a key predictor of motorcycle crashes.

Many behaviors that are dangerous behind the wheel of a car are more dangerous on a motorcycle. In the general population, drunk driving and speeding are responsible for greater percentages of crashes on motorcycles than in cars, and close to half of single-vehicle fatal motorcycle crashes are due to drinking. Evidence shows that the ability to handle a motorcycle begins declining before the rider reaches the legal BAC limit of 0.08. Service Safety Center briefs to PMV TF have indicated that drinking-related motorcycle mishaps appear to be lower in the military sport-bike population than among their civilian counterparts, but this conclusion appears to be anecdotal at this point.

Several other factors also pay a role in motorcycle crashes. Lack of helmet use is associated with about half of fatal crashes—a similar percentage to the association between car crashes and lack of seat belt wearing, but fewer motorcyclists wear helmets than car occupants wear seat belts. Motorcyclists without licenses are more likely to crash than those who have them, and better conspicuity of both rider and bike seems to reduce crashes since riders are more visible to cars.

Not all drivers and motorcyclists are equally likely to take risks. Young men engage more often in risky behaviors than older drivers, but there are other ways to identify more specifically those who are likely to be risky drivers. Within both the military and the general population, those who drink heavily are more likely to drive after drinking than those who are light drinkers. This is a particular problem for the military, which has higher rates of heavy drinking than comparable civilian populations. Those rates have also been rising over the past decade. People who have previously driven drunk or been in crashes are more likely to do so again than people who have not.

Certain personality types have also been found to take more risks. Sensation-seekers thrive on exciting experiences, including speeding. Some researchers have theorized that sensation-seekers speed because they love going fast, not because they don't understand the risk. People who are impulsive take actions without heeding the risk. The military has far higher percentages of people who score high on sensation-seeking and impulsiveness than the general population.

Finally, certain beliefs seem to contribute to the propensity to take risks. There is some limited evidence that people who view certain driving behavior to be less risky are more willing to take those risks. People who are less concerned with social norms regarding safe driving, have higher perceptions of their own driving skill, and worry less about the consequences of unsafe driving tend to take more risks. And in one interesting study, young men who were encouraged to think "macho" thoughts tended to drive more recklessly.

Safety Interventions

Because traffic safety has been a concern since Americans first started driving, many policies have been adopted to encourage safer driving. One method has been media campaigns, which spread a particular message in an attempt to get drivers to change their behavior. Some of these, particularly with respect to drunk driving, have been effective, although it is difficult to separate the effects of specific campaigns from larger changes in people's attitudes. Media campaigns work best when exposure is broad, messages are targeted to the audience, the environment supports the change in behavior, the campaigns are based on theory, and the campaign effectiveness can be analyzed. A carefully planned and executed media campaign includes basing the campaign in communication theory, pretesting the campaign to ensure it works with the targeted audience, and ensuring that the message is strategically placed where it reaches that audience repeatedly.

Many campaigns use an approach called a "fear appeal," which tries to scare drivers into changing their behavior by emphasizing the risks and consequences of such behaviors as drinking and driving or speeding. While fear appeals can be effective, one problem is that they tend to backfire when the intended audience is sensation-seeking. An approach called SENTAR (sensation targeting) is relatively new but has been successfully used with sensation-seeking teenagers in an antidrug campaign.

High-visibility enforcement campaigns combine stepped-up enforcement of laws with publicity about enforcement. This method has been quite successful with both drunk driving campaigns and seat belt campaigns, including some applications on military bases.

Several safety measures have shown promise with respect to drinking and driving. Forty-one states have automatic license suspension, under which drivers who fail or refuse breath analysis testing can have their licenses suspended by the police officers who stopped them. Alcohol treatment in general can reduce drunk driving, with the added benefit of identifying other mental health problems (such as posttraumatic stress disorder or depression) that often accompany alcoholism. Finally, teaching bartenders to refuse service to intoxicated patrons has been shown to be effective in a military context.

With respect to initiatives specifically for motorcyclists, helmet laws—which the military already has in place—have been found to reduce fatalities in states where all riders are required to wear helmets. States with partial helmet laws (meaning that only riders under a certain age are required to wear helmets) have fatality rates similar to states with no laws, since partial helmet laws are difficult to enforce and many riders who are legally required to wear helmets do not.

Two other policies—rider training and various types of graduated licensing—have shown mixed results. Rider training has been extensively studied but without reaching any conclusive assessment. Some research has found that training reduces the frequency of crashes, while other research has found no effect or even an uptick in crashes. It is possible that the research has not been designed well enough to eliminate the possibility that riders who voluntarily take training are more motivated to ride safely than those who take it because it is mandatory, or it is possible that the motorcyclists' attitudes about risk-taking are more important than the training.

Study results were also mixed for graduated licensing. Three forms of graduated licensing all provide for some interval and training between starting to ride and getting a full license, or allow the motorcyclist to ride only certain types of motorcycles. Learners' permits allow riding only under supervision, graduated drivers' license programs have a restricted license between the learner's permit and a full license, and tiered licenses allow motorcyclists to ride only a specific type of bike. One study found that certain types of learners' permits reduced fatalities, but that graduated drivers' licenses and tiered licenses had no effect. Studies of graduated drivers' licenses—which are in widespread use for driving a car but are less common for motorcycles—have found that they can be effective by reducing the amount of motorcycle riding.

Finally, protective clothing for motorcyclists is effective in reducing certain types of injuries in lower-speed mishaps—generally abrasions and other soft tissue injuries—but not fractures and other injuries generally seen as a result of ejection-related blunt-force trauma. Helmets remain the most important piece of protective gear.

Findings

This review shows that the following safety interventions, which are not listed in any particular order, tend to help in the reduction of vehicle crashes.

- Better enforcement of underage drinking laws and continuation of alcohol deglamorization campaigns. DoD regulations exist, but underage drinking seems to be relatively common.
- High-visibility enforcement techniques for sobriety checkpoints.
- High-visibility enforcement techniques for seat belt use.
- Adoption of a lower BAC level (such as 0.05) for motorcyclists, since the evidence shows that motorcyclists' ability to drive safely begins declining at lower BAC levels than those for car drivers.
- Screening—perhaps as part of a medical assessment—and brief intervention with a trained counselor for at-risk drinkers, since they are at higher risk for drinking and driving.
- Media campaigns that are paired with community activities that also emphasize driver safety, such as workshops or fairs and with enforcement of driving regulations, and targeted at the drivers at highest risk (men in their teens and early 20s).
- Requirements that motorcyclists be licensed and own their vehicles. Enforcement of those requirements means that motorcyclists found to be lacking a valid license or to be borrowing vehicles would be punished.

This review did not identify any safety interventions that had been effective specifically in reducing speeding, distracted driving, or fatigued driving.

Suggestions for Further Research

The military may wish to conduct additional targeted research in six areas:

- mapping the locations and researching the causes of crashes among military members
- assessing in more detail and in combination the factors that seem to be particularly problematic for military populations, such as deployment, stress, and sensation-seeking
- determining the effectiveness of various types of motorcycle training on rider skill and risk-taking
- assessing ongoing military-specific safety interventions
- developing a systematic service-wide approach to identify the motorcycle rider population, since it is possible that the risk profile of motorcyclists is different than that of car drivers
- determining whether fatigued and distracted driving are major predictors of vehicle crashes for military members.

Acknowledgments

This work was completed with the assistance of DSOC. Joseph Angello, in the Office of the Under Secretary of Defense for Readiness, serves as the DSOC executive secretary. Air Force Maj Gen Fred Roggero chairs PMV TF. In PMV TF, Air Force Col Don White serves as deputy chair, Air Force Majs Dan Roberts and Brian Musselman serve as the executive secretaries. Robert A. Gardiner of NDCEE for the Office of the Deputy Under Secretary of Defense for Readiness (Readiness Programming and Assessment) provides DSOC staff support and task force coordination.

The task force includes representatives from the U.S. Air Force Safety Center (Col Roberto Guerrero and Bob Baker); U.S. Army Combat Readiness Center (Walter Beckman and Earnest Randle); U.S. Naval Safety Center (CAPT Steve Johnson and Donald Borkoski); U.S. Marine Corps Safety Center (Cols. Paul Fortunato and James Grace, Peter Hill, and John Waltman), and U.S. Coast Guard Safety (Leslie Holland and Dale Wisnieski). Other task force members and partners include the following advisors: John Seibert, from the Office of the Secretary of Defense, Environment, Safety, and Occupational Health; Kurt Garbow, from the Office of the Deputy Assistant Secretary of the Navy (Safety); operational leaders from each of the services; Office of the Secretary of Defense DSOC Program Manager Jerry Aslinger; NDCEE DSOC Program Manager Karen F. Nelson; NDCEE epidemiologists and data analysts Melissa Hartsell-Riester, Tom Vincent, and Kerry Campbell; Debra Ann Ferris of the National Safety Council; Charlie Fernandez of the Motorcycle Safety Federation, and Keith Williams of the U.S. Department of Transportation. We appreciate the support of DSOC and the entire task force in providing data and publications that helped us produce this report.

The research team thanks Catherine Piacente for her assistance preparing the document and references and Alison Raab Labonte for conducting some of the initial literature searches. We also thank reviewers David Loughran, senior economist at RAND, and Sheila Mitra-Sarkar, director of the California Institute of Transportation Safety at San Diego State University, for their thoughtful suggestions.

Abbreviations

ADMP	active-duty military personnel
ASMIS	U.S. Army Safety Management Information System
BAC	blood alcohol concentration
BTS	Bureau of Transportation Statistics
cc	cubic centimeters
CDC	Centers for Disease Control and Prevention
CIOT	click it or ticket
DoD	Department of Defense
DOT	Department of Transportation
DRLs	daytime running lights
DSOC	Defense Safety Oversight Council
DUI	driving under the influence
DWI	driving while intoxicated
FARS	Fatality Analysis Reporting System
FMVSS	Federal Motor Vehicle Safety Standard
FY	fiscal year
g/dL	grams per decaliter
GDL	graduated driver licensing
HRBS	Department of Defense Survey of Health Related Behaviors Among Active Duty Military Personnel
HVE	high-visibility enforcement
IIHS	Insurance Institute for Highway Safety
MIC	Motorcycle Industry Council
MSF	Motorcycle Safety Foundation
NCSA	National Center for Statistics and Analysis
NDCEE	National Defense Center for Energy and Environment
NHTSA	National Highway Traffic Safety Administration
OEF/OIF	Operation Enduring Freedom/Operation Iraqi Freedom
PCP	phencyclidine

PMV	Private Motor Vehicle
PPE	personal protective equipment
PTSD	posttraumatic stress disorder
SENTAR	sensation targeting
TF	Task Force
VMT	vehicle miles traveled

CHAPTER ONE

Introduction

Deaths from vehicle crashes are declining in the United States, but crashes still claim the lives of over 35,000 Americans each year (National Highway Traffic Safety Administration [NHTSA], 2009c). Traffic safety has been the focus of intense research over many years, with researchers examining topics ranging from the role of vehicle safety features and road design to human factors and the psychology of risk-taking behind the wheel. While this understanding has contributed to making driving safer, much progress can still be made.

For the U.S. military, deaths from vehicle crashes are a key safety concern. The population that dominates the military—young men—is the group most at risk from dying in a vehicle crash. Data from the military have shown that vehicle crashes are the leading cause of death in peacetime (Powell et al., 2000), and the loss of life and injuries contribute to readiness and training costs.

The military faces several issues with respect to vehicle crashes. First, the young male demographic is at the highest risk from vehicle crashes, so the number of people killed by crashes in the services per 100,000 population is generally higher than that for the U.S. population, although rates are comparable for civilian men of similar ages. Second, a large percentage of deaths occur on motorcycles, and motorcycles are the one area in which data show that riding is getting more dangerous, rather than safer, over time.

This research was undertaken for the Private Motor Vehicle (PMV) Task Force (TF), one of a number of task forces under the Defense Oversight Safety Council. PMV TF includes operational leaders, traffic safety officers, and certified safety professionals drawn from across the services who seek to reduce vehicle crash deaths and injuries through drafting and enforcing traffic and safety regulations, providing training, and developing and implementing safety campaigns. This report provides the task force with background on how military fatalities compare with those in the civilian population, as well as a review of empirical studies that can contribute to the understanding of what is known about vehicle crashes, how to prevent them, and how to reduce their severity.

Our goal was to provide suggestions to reduce preventable crashes in the military based on what researchers know about who is most at risk and which safety interventions are most effective. There are two reasons why most of our suggested policies are based on the civilian literature on risky driving and safety interventions. First, we found a fairly small number of studies that discuss crash risks for military members; these are discussed extensively in Chapter Three. What these studies concluded about risk for military members is similar to that for civilians: Men are at greater risk than women, young men at greater risk than older men, those who drink heavily at greater risk than those who do not, and so forth. So we have drawn conclusions from the civilian literature that apply most strongly to a military context: issues of risk-taking, drunk driving, and motorcycle safety. Second, some of the data we have available in the civilian world with regard to driving behavior are not available for military personnel. For example, U.S. surveys conducted periodically show the estimated number of miles driven by different demographic groups, but we are not aware of any military-specific data that show the average number of miles driven. Without such data, we do not know if military members are at greater, lesser, or similar risk for crashes than their civilian counterparts per mile driven.

The team did not undertake new research for this study. We did collect a modest amount of data on vehicle crash deaths among those in the various services and have presented it in a descriptive fashion. Most of the report relies on evidence-based studies that have been published in the academic literature and on some reports that, while perhaps not peer reviewed, were issued by U.S. government agencies or other reputable organizations with a history of working in traffic safety. In the course of this work, we collected a library of more than 500 citations, of which over 200 are cited in this report. Some of these publications were provided by the task force; others were located through searches in a number of databases or were found because they were referenced in other documents. Not every report we reviewed is included here; we discuss those that were of particular relevance, because they (1) looked at military populations or factors that are prevalent in the military, such as high sensation-seeking; (2) summarized other research on a specific topic; and/or (3) were, in our judgment, of high methodological rigor. Space does not permit us to discuss each report we reviewed in depth; rather, this report focuses on key findings across the literature.

In several areas of inquiry of interest to the task force, we found very little empirical evidence. For example, we found very little work on combat driving behaviors that influence personal driving; the one study we located was still in progress (Stern and Rockwood, forthcoming). With regard to other topics, such as fatigue and distracted driving, we did not identify any studies specific to U.S. military personnel, but we have included studies of military personnel in other countries and among young drivers because they are relevant to the task force.

The charge of the PMV TF is to reduce preventable vehicle crashes, so our main focus was on demographics and driver behavior. The team assessed demographic factors to guide the task force in developing safety campaigns that target those most at risk. We looked at driver behavior as it contributes to crashes in order to help guide campaigns to reduce the incidence of the most dangerous behaviors behind the wheel. The focus on preventable crashes means that we did not look at factors outside the control of the driver, such as weather and road conditions. With regard to passenger vehicles, we did not examine vehicle features, such as the presence of air bags, that may affect survivability in a crash. We also did not look at vehicle performance problems, such as brake failure, that may cause crashes.

With respect to motorcycles, we expanded the scope of the study to some extent, since the task force has a particular interest in motorcycle safety and the selection of the bike itself may make a substantial difference in safety outcomes. So we present data linking engine size and crash rates, as well as evidence on the effectiveness of other features, such as daytime running lights (DRLs). These may help guide suggestions about safety campaigns regarding issues specific to motorcycle safety that are not relevant for passenger cars.

The report broadly addresses two themes: who is at risk and what can be done to reduce that risk. The first theme includes what is known about which demographic groups are at highest risk, which driver behaviors are considered dangerous, and what is known about personality and risk-taking. The second presents findings from the literature about which safety campaigns and regulations have been effective and makes suggestions to the task force. In both of these areas, we have split out motorcycles into a separate chapter because some of the risk factors are different and the issues about regulations and vehicle and equipment selection are distinct from those of automobiles. Most of the evidence as it relates to demographics, driver behavior, and risk factors is based on all vehicle crashes, not passenger vehicles specifically. Therefore, several chapters focus exclusively on motorcycles.

Including this introduction, the report contains eight chapters. Chapter Two provides context for the report, showing broad trends in vehicle fatalities among the U.S. population and military service members. Chapter Three presents factors that influence who is most likely to be in vehicle crashes, both in demographic terms as well as with regard to driver behavior. This chapter also discusses several studies specific to military personnel. Chapter Four is similar to Chapter Three, but with a particular focus on motorcycles, since many of the risk factors for motorcyclists differ from those of car drivers. Chapter Five examines the literature on risk-taking and personality types of those most likely to take risks behind the wheel. Chapter Six reviews the evidence from driving safety and enforcement campaigns with regard to which techniques have been shown to be effective, especially among young, male, and risk-taking populations. Chapter Seven reviews the evidence about safety gear, training, and other policies specific to motorcyclists. Chapter Eight concludes with suggested policies for the task force's consideration based on the evidence reviewed for this report.

CHAPTER TWO

Deaths from Motor Vehicle Crashes

In this chapter, we present general trends in motor vehicle crash rates for the United States as a whole and for its military population. For this comparison, we examined fatalities from motor vehicle crashes. There are four reasons for concentrating on fatalities:

1. While injury crashes can certainly be very serious, the range of possible injuries is quite wide, and this makes it difficult to compare data across sources. In some databases, it can be difficult or impossible to distinguish between serious injuries and relatively minor ones. Data on fatalities are more readily available and more easily compared across data sources.
2. We are not confident that all injury crashes are reported to authorities, because there may be reasons that drivers or passengers choose not to report them (for example, if they are uninsured). Therefore, the databases that measure injuries may be incomplete. However, well-developed systems are in place for reporting fatalities, so this database is more accurate.
3. Fatalities obviously represent the worst possible outcome to a crash and thus serve as a good measure of how much safety has changed over time.
4. Although this report is ultimately concerned with reducing all types of crashes, the literature review suggests that factors contributing to fatal crashes also contribute to injury crashes. Identifying the factors associated with fatal crashes and developing strategies to decrease them should decrease injury crashes as well.

However, we also include data on injuries as available and appropriate. Unless otherwise defined more precisely, *injury* refers to any type of injury, regardless of severity.[1] An *injury crash* is a crash that causes any type of injury.

By several measures, driving a car is becoming progressively less dangerous over time, while riding a motorcycle is becoming increasingly dangerous. This chapter looks at some of the components of those trends.

Overall Rates

Deaths from crashes can be measured in several ways; this chapter uses three of the most common measures. *Absolute numbers* of fatalities is a term for the total numbers of persons

[1] Generally, injuries can be tracked only if a person involved in a crash reports any injury to the police or seeks medical attention. Therefore, if neither of these occurs, some very minor injuries may not be included in data sets.

killed in vehicle crashes. *Deaths per vehicle miles traveled* (VMT) is a term that assesses the relative risk of death against the amount of driving in order to assess the relative safety of driving. Finally, *death rates per 100,000 population* is a term that compares the number of deaths with the number of persons in a particular demographic group in order to determine relative incidence. Such rates can then be used to compare the prevalence of a cause of death across different groups of people. In this chapter, we will refer to different measurements depending on available data.

One weakness of this report is that the available data do not necessarily describe the issues that are most relevant to the military. Of the three measures above, the one of most interest to the military is fatalities per VMT, because this captures the risk of driving. This measure depends in turn on two other ratios: the number of crashes per mile driven (crashes per VMT) and the chance that a crash will be fatal (fatalities per crash). However, given the limitations of the available data, we are not able to calculate either of these ratios for service members. We cannot even calculate fatalities per VMT, because we do not know how many miles are driven by military personnel.

Therefore, the best way we have to analyze the prevalence of military vehicle crashes is fatalities per 100,000 service member population. While this is not an ideal method to describe the risk of driving, it at least provides a way to look at trends over time and to compare military fatalities to civilian ones. Since the nature of this project is a review of available literature, we cannot answer the question of how many crashes military personnel have per mile driven or whether crashes are similarly lethal for civilians and military personnel of similar demographics.

Some statistics distinguish between the three categories of people who can be injured or killed in a crash—drivers, passengers, and those outside the vehicle (such as pedestrians or bicyclists). Drivers and passengers are both considered vehicle occupants. We will make these distinctions as appropriate (for example, the use of seat belts does not affect whether pedestrians are killed).

This chapter draws heavily on two sources of fatality data. NHTSA collects and reports fatality data in the Fatality Analysis Reporting System (FARS). These data include all fatalities that result within 30 days of a vehicle crash. FARS data provide information about the location and circumstances of the crash, the types of vehicles, and the people involved in the fatal crashes, such as their age, sex, previous violations charged, and home zip codes.[2] The Centers for Disease Control and Prevention (CDC) also publish mortality data in their series of National Vital Statistics publications (formerly called the Monthly Vital Statistics reports). The CDC data can be used to track death rates by specific demographic groups. The two sets of statistics vary slightly; for example, CDC includes deaths in the year that the death occurred, and it does not limit crash fatalities to a period of 30 days, as the FARS data do. However, the trends shown in the statistics are very similar.[3] Our period of analysis is generally since 1991, because data were readily available in this time frame.

---

[2] FARS data about people involved in fatal crashes are limited to information found on police accident reports and motor vehicle records.

[3] The CDC data show consistently higher rates of death than the FARS data, perhaps because they are adjusted by age, whereas the FARS data are not.

Total Crash Deaths

Absolute numbers of fatal crashes declined slightly in the early 1990s and have remained relatively constant since. From 1992 to 2007, the total number of Americans killed in vehicle crashes has been between 39,250 and 43,500 annually (Bureau of Transportation Statistics [BTS], 2009). These numbers saw a sharp decline in 2008, to about 37,000. NHTSA predicts that 2009 fatalities will fall even further, based on analysis of partial 2009 data (National Center for Statistics and Analysis [NCSA], 2009b). While there has been no definitive analysis of what has caused the recent decline, some observers have speculated that it is linked with a decline in the amount that Americans drive, driving at slightly slower speeds, and the effects of a variety of safety measures adopted over the years (White, 2009).

CDC's most recent report on causes of death shows that accidents (of all types) are the fifth leading cause of death in the United States. In 2006 (the most recent year for which full CDC statistics are available), motor vehicle crashes represented just over one-third of all accidents (45,300 of 121,600 accidental deaths). However, deaths from motor vehicle crashes are not evenly distributed among the population. They represent a greater proportion of deaths among men than women, and among people up to their 30s than older adults. A detailed analysis of 2002 CDC data showed, for example, that for men, crashes (excluding other accident types) were the sixth leading cause of death, but the tenth leading cause for women. For males, motor vehicle crashes were the leading cause of death from age 3 to 34. For females, they were the leading cause of death from age 6 to 29 (Subramanian, 2005).

Fatal Crashes per Vehicle Mile Traveled

Fatal crashes per VMT measures the number of crashes in relation to how much people drive. This measurement probably provides the best assessment of risk, since if crash fatalities decline while the number of VMT increases, that means that overall, driving is becoming safer.

In 1990, the rate was 2.08 fatalities per 100 million VMT; in 2008, it was 1.27 (BTS, 2009). This slight but steady decline occurred despite generally increasing VMT per capita. From 1991 to 2008, VMT per capita increased from about 8,500 to 9,500 (Puentes and Tomer, 2008), although 2008 and 2009 both show declines in VMT over the previous year. These figures show that driving has continued to become safer; as the American population both increases and drives more, the number of fatal crashes stays about the same. This result could be because of either lower crash risk or the fact that crashes are less likely to be fatal (for example, because of improvements in medical care or better vehicle safety features). Figure 2.1 shows the declines in total fatalities as well as in deaths per 100 million VMT.

Figure 2.2 shows the number of persons killed per VMT by age and gender in 2001.[4] This shows that men in their teens and early 20s are the group at highest risk for being killed in a vehicle crash per mile driven. While rates for females begin falling in their late teens, rates for men continue to rise. This would seem to suggest that factors beyond driving experience are at work. From about age 25 to 60, men are at greater risk for being killed in a crash than women, even when we control for the number of miles driven. Note that the number of persons killed in vehicle crashes is different than the number of drivers who caused those crashes,

[4] The reason for not using more-recent data is that the survey that collects information on the VMT driven by different demographic groups—the National Household Travel Survey—is not conducted every year. It was last conducted in 2009, but vehicle crash data for 2009 were not available when this report was written. The next most recent survey dates from 2001.

Figure 2.1
Total Deaths from Vehicle Crashes and Deaths per 100 Million VMT, 1990 to 2008

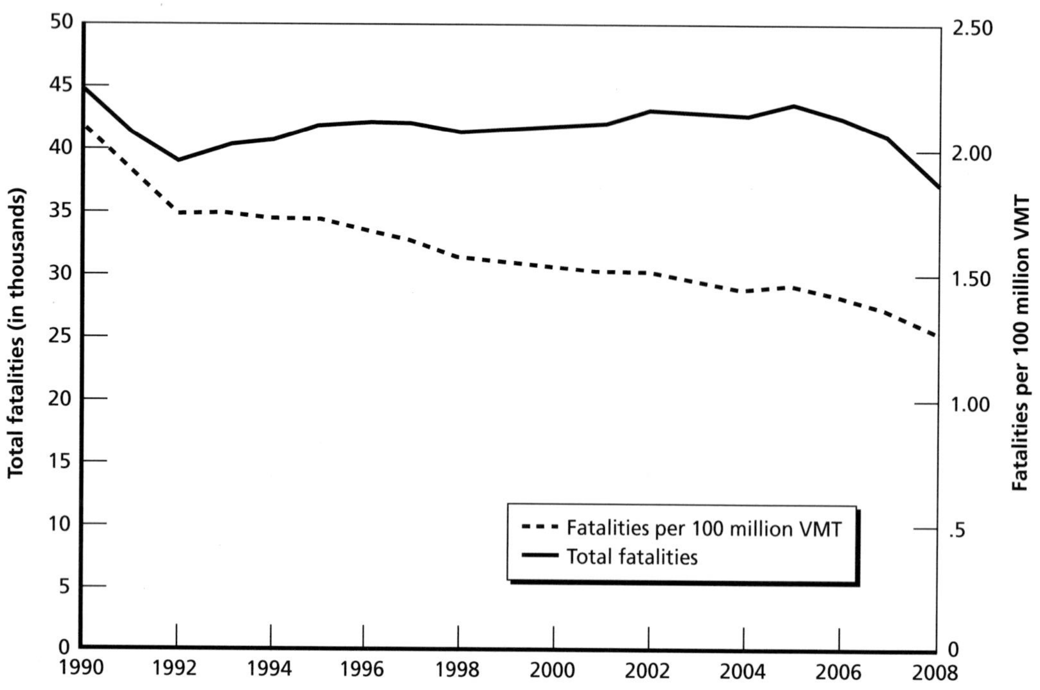

SOURCE: BTS, 2009, Table 2-17.
RAND *TR820-2.1*

since some persons killed are passengers, bicyclists, or pedestrians. However, as discussed in Chapter Three, young men are generally the riskiest drivers of any group.

We would like to be able to compare crash deaths per VMT for young male civilians and the military, to determine whether military personnel are at greater or lesser risk per mile driven than their civilian counterparts. However, as noted earlier, VMT data for the military are not available, so we cannot make this comparison.

Crash Deaths per 100,000 Population

FARS data indicate that death rates from motor vehicle crashes have declined over the past two decades. The overall (unadjusted) rate fell from 16.5 deaths per 100,000 population in 1991 to 12.3 in 2008 (NHTSA, 2009c).

Death rates from motor vehicle crashes are particularly high among teenaged boys and young men, although they have been declining. From 1991 to 2006, the last year for which full CDC data are available, the death rate for men aged 15 to 24 has fallen from just over 45 per 100,000 population in 1991 to about 36 per 100,000 population.

U.S. Military Crash Deaths

The best direct comparison of risk we can make between the civilian and military population is crash fatalities per 100,000 population. Within the military, deaths from motor vehicle crashes are higher than the U.S. average, but somewhat lower than for the male populations of

Figure 2.2
Vehicle Crash Deaths per 100 Million VMT by Age and Gender, 2001

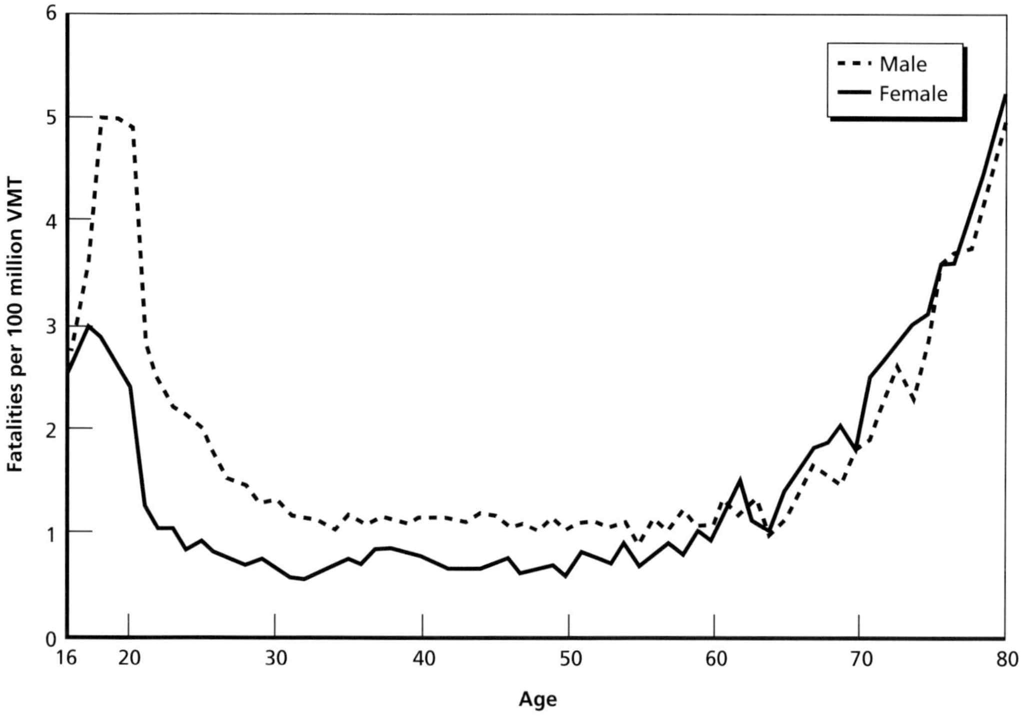

SOURCES: VMT data are from Federal Highway Administration, 2001; data on vehicle crash deaths are from NHTSA, 2009c; and population estimates are from CDC, 2003.
RAND *TR820-2.2*

similar ages. During peacetime, crashes are the leading cause of death of military personnel; from 1982 to 1992, for example, motor vehicle fatalities accounted for 20 to 40 percent of all military fatalities, the largest single cause of death (Hooper et al., 2005). Our own analysis, based on data provided by the services and total number of accidents from the Defense Manpower Data Center, shows that since 2000, about 40 to 55 percent of accidental deaths in the military are due to vehicle crashes. Table 2.1 shows selected U.S. population death rates from vehicle crashes, and Table 2.2 shows the ten-year average for fatalities per 100,000 population for the Marines, Coast Guard, Army, Navy, and Air Force.[5]

We have shown the ten-year average, rather than the rate for each year, because the rates in the services are more volatile on an annual basis since the populations are smaller; a difference of a small number of fatalities results in a substantial change in the rate. It is difficult to discern from these data (available in the Appendix) whether there are important overall trends. While all five services show volatile rates on a year-over-year basis, the Marine Corps rate is consistently higher and the Air Force rate the lowest. These figures have not been adjusted to account for differences in demographics between the services.

Although this comparison is reasonably straightforward, it is difficult to ascertain how risk for military personnel differs from civilians of similar ages and gender. First, the military crash deaths are not adjusted for demographics by service, so if the services have different pro-

[5] Each service compiles its own statistics, and there may be differences in how they are gathered and analyzed.

Table 2.1
Death Rates from Vehicle Crashes for the U.S. Population, 2000 to 2006

	Seven-Year Average	Highest Rate	Lowest Rate
U.S. men 15–24	37.3	39.3	36.4
U.S. men 24–35	24.1	24.5	23.7
Total in the United States	14.7	15.0	14.3

SOURCE: CDC, 2009.

Table 2.2
Death Rates from Vehicle Crashes for the Military Services, 2000 to 2009

	Ten-Year Average	Highest Rate	Lowest Rate
Marines	27.1	35.3	19.0
Coast Guard	19.6	25.0	9.5
Army	17.7	21.0	15.0
Navy	15.9	20.3	9.9
Air Force	11.9	17.2	7.5

SOURCES: Navy and Marines: Naval Safety Center, 2010; Army: ASMIS, 2009; Air Force: Terry L. Todd, Air Force Senior Master Sergeant, email to Liisa Ecola on January 4, 2010; Coast Guard: Dale Wisnieski of the U.S. Coast Guard, email to Liisa Ecola on February 17, 2010.

portions of women or older members—who have lower crash risks than young men—these are not apples-to-apples comparisons. Second, these figures are not adjusted for VMT, because VMT for service members are not available. If service members drive considerably less than civilians, perhaps because of living on a base, but are in similar numbers of fatal crashes per population group, that would mean that they face increased risks from driving. However, we are not able to say if this supposition is true. We also do not have information on the proportion of crashes that take place on bases as opposed to off base.

Motorcycle Crash Deaths

This section looks specifically at deaths from motorcycle crashes among civilian and military populations. In contrast to overall crash deaths, which have declined and held steady over the past several decades, motorcycle crash deaths dropped from the late 1970s through the mid-1990s and have been rising dramatically since.

While total motorcycle fatalities are almost 15 percent of all crash fatalities, riding a motorcycle is much more dangerous on a per-mile basis than driving a four-wheeled vehicle. Figure 2.3 shows the total fatalities and the fatalities per VMT for all vehicle crashes and for motorcycles. The results are sobering; as NHTSA noted, "Per vehicle mile traveled in 2007, motorcyclists were about 37 times more likely than passenger car occupants to die in a motor vehicle traffic crash and 9 times more likely to be injured" (NHTSA, 2009a, p. 3).

Figure 2.3
Fatalities per 100 Million VMT for All Vehicles and Motorcycles, 1992 to 2008

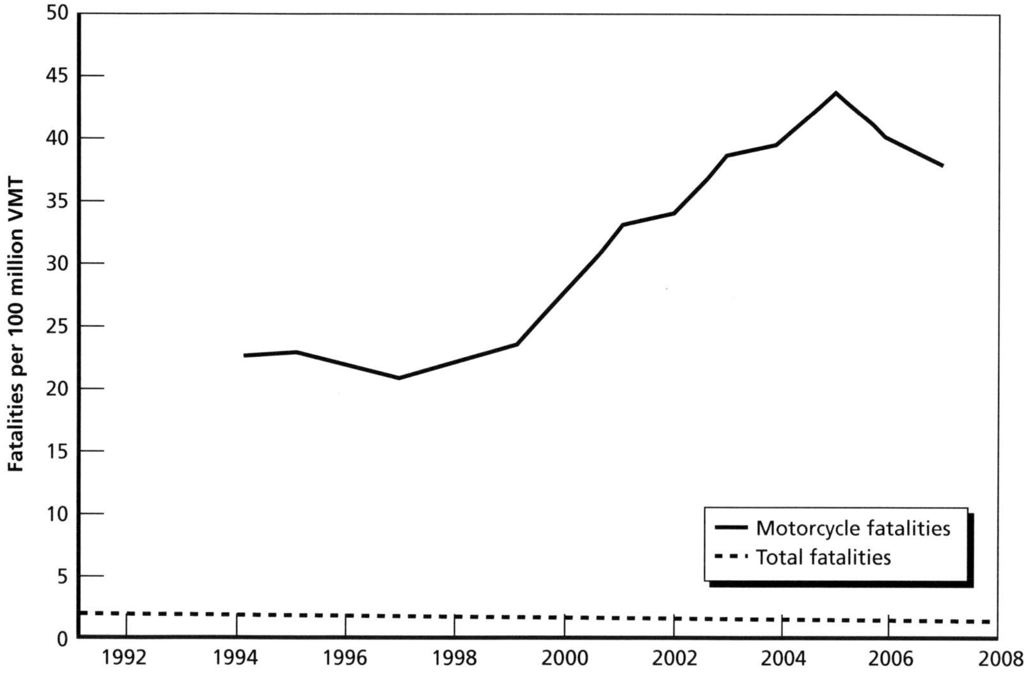

Fatalities per 100 million VMT
50
45
40
35
30
25
20
15
10
5
0
1992 1994 1996 1998 2000 2002 2004 2006 2008
Motorcycle fatalities
Total fatalities

SOURCE: BTS, 2009, Appendix D.
RAND *TR820-2.3*

While the number of fatal crashes closely tracks the number of motorcycles registered, the risk of crashing on a motorcycle is increasing. Over the past decade, the number of registered motorcycles and fatalities has increased while the VMT on motorcycles has remained generally constant, meaning that mile for mile, riding a motorcycle is much riskier now than it was ten years ago (as Figure 2.3 demonstrates). While these rates have been rising among all types of riders and motorcycles, some patterns have emerged as to who faces the greatest risks; these will be discussed in detail in Chapter Four.

It is difficult to compare death rates from motorcycle crashes in the military with those in the civilian population without accurate numbers of motorcycle riders or VMT on motorcycles by service members. For example, it is possible that the military has a much higher share of motorcycle riders than the U.S. population, or that motorcyclists in the military are more apt to ride motorcycles instead of cars, while civilian motorists may ride motorcycles only for pleasure and thus drive most of their miles in different circumstances.

Figure 2.4 shows how the proportion of vehicle deaths that occurred on motorcycles among military members and the total U.S. population has changed since 2001. We selected the year 2001 because the four services[6] had similar rates that year: between 15 and 20 percent of crashes occurred on motorcycles. Among the U.S. population, the proportion of motor vehicle deaths occurring on motorcycles has increased from about 7.5 percent in 2001 to almost 15 percent in 2009, meaning it has nearly doubled. For the military services, this percentage has

[6] The data series provided by the Coast Guard began in 2003, so we could not compare growth in the proportion of motorcycle crashes to the proportion in the 2001 base year.

Figure 2.4
Change Since 2001 in Percentage of Fatalities on Motorcycles

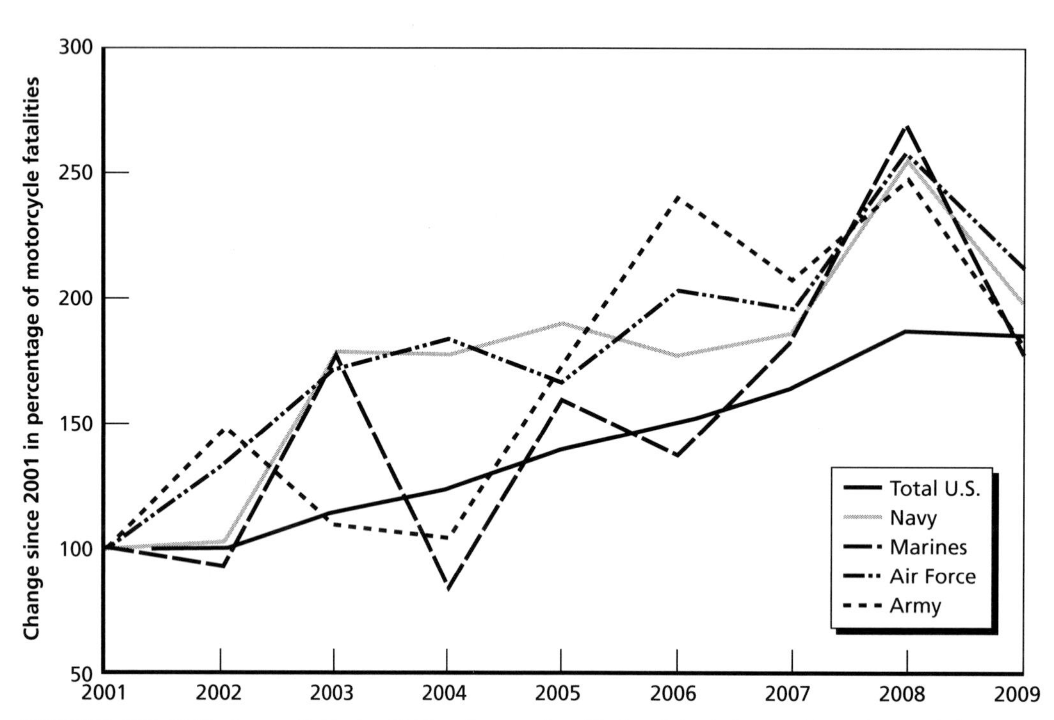

SOURCES: BTS, 2009, Appendix D; the 2009 estimate of traffic fatalities is from NCSA, 2010; and the 2009 estimate of motorcycle fatalities is from Hedlund, 2010. Navy and Marine estimates are from Naval Safety Center, 2009. Army estimates are from Walter Beckman of U.S. Army Combat Readiness/Safety Center, email to Liisa Ecola on February 16, 2010. Air Force estimates are from Terry L. Todd, Air Force Senior Master Sergeant, email to Liisa Ecola on January 4, 2010.
NOTE: Coast Guard data for the percentage of fatalities on motorcycles were not included here because the data were available only from 2003 onward.
RAND *TR820-2.4*

risen more quickly, increasing by 2.5 times through 2008, although all the figures among the services as well as the total U.S. figure show a decline for 2009. While the U.S. total includes crashes in all demographic groups and is not directly comparable to military populations, it would be complex to do a direct demographic comparison because of the lack of information on motorcyclists in the U.S. population. In the Navy, Marines, and Coast Guard, the proportion of motor vehicle deaths occurring on motorcycles has been 50 percent or more some years. The Coast Guard generally has the highest proportion of motorcycle deaths, but the total numbers are quite small (three fatalities in fiscal year [FY] 2009), so one crash can have a large effect on the rate.

CHAPTER THREE

Factors Influencing Crash Rates for All Vehicles

Why are some people more likely to be killed in a motor vehicle crash than others? Research into traffic safety looks at two broad areas that help shed light on this question: the occurrence of crashes and their severity. The factors discussed in this chapter influence one or both of those areas. For example, wearing a seat belt does not generally influence whether a crash occurs, but it does influence the severity.

Because the PMV TF focus is directed at reducing "preventable" vehicle crashes, this chapter will discuss demographic and behavioral factors as they relate to all vehicles (factors specific to motorcycles are covered in Chapter Four). Obviously many other factors influence the likelihood and severity of crashes—vehicle features such as air bags, roadway features such as sharp turns, weather, etc. We have chosen to look into those elements that the military is able to influence; certain segments of the population can be targeted for campaigns, and behavior can be changed. Some policymakers and researchers take a broader view than others as to what is "preventable"—Sweden has set a goal of no fatal vehicle crashes[1]—but in this analysis, we assume that road conditions, vehicle safety features, and other such elements are beyond the task force's purview.

Given the many possible causes of motor vehicle crashes, it is difficult to isolate the effect of any specific cause. With regard to a single crash, many factors may come into play: the driver's behaviors, the condition of the road and vehicle, weather and visibility, and so forth. Since crashes may have multiple causes, a fatality may be attributed to both excessive speed and not wearing a seat belt.[2]

Furthermore, the chance that any given trip will result in a fatal crash is very low. Much of the research we review, therefore, looks at relative risk; that is, how much more dangerous driving under a certain condition may be than in the absence of that condition. Studies of crashes tend to analyze the contributions of a particular factor, based largely on information available in FARS or other databases.

While many of these factors apply to both civilian and military populations, this chapter also reviews what is known about factors specific to the military. Some of the military studies had access to information about members' behavior before they crashed, such as whether some-

[1] The Swedish National Road Administration's official policy is "Vision Zero," a goal of no fatalities or serious injuries from vehicle crashes. It was adopted by the Swedish Parliament in October 1997. See discussions in Elvik (1999) and Rosencrantz (2006) about the policy's feasibility and effectiveness.

[2] This is why the proportion of crashes attributed to particular causes does not sum to 100.

body drank heavily, that allowed these studies to create models using many variables to try to determine the relative odds of being in a crash.[3]

General Demographic Factors

As Figure 2.2 clearly showed, age and gender are strongly associated with crash rates. Once teenagers reach driving age, men are roughly twice as likely to be killed in a motor vehicle crash than women. This differential is highest for young drivers and generally constant through the working years, then the gap closes in the mid-60s. While death rates vary slightly by race and ethnicity, those factors are far less important than gender and age. For example, in 2006, the age-adjusted crash death rate for white males was 21.8, while for black males it was 22.6 and for Hispanic males, 21.2 (Heron et al., 2009).

Some, but not all, of the disparity may result from men driving more miles than women. Yet women experience only 0.7 deaths per 100 million VMT, while men experience 1.3 (Gerard et al., 2007). A likely explanation for the difference in deaths per VMT is the greater likelihood that men engage in risky behavior while driving, as we discuss below (Vanderbilt, 2008).

Young drivers are more likely than adults over 25 to be killed in motor vehicle crashes. For teenaged drivers, lack of experience, more risky behavior (such as speeding and not wearing seat belts), lower perceptions of risk, and higher rates of fatigue all contribute to a greater likelihood of crashes. While driving at night and drinking increase all drivers' risk, they do so even more for teenagers. Finally, driving with passengers actually increases teenaged drivers' crash risk, while driving with passengers reduces the risk for adults (Shope and Bingham, 2008). While one recent paper explores a potential link between teens living in poverty and higher crash rates, the county-level analysis did not look at the impact of the differences in urban versus rural driving, and the counties with the highest poverty rates were also heavily rural (Males, 2009).

General Behavioral Factors

As mentioned above, demographics play a role in crash rates largely because they are associated with drivers' behaviors. This section looks at some driver behaviors that increase the risk of crashes and, generally, in what ways or by how much. Examined below are the effects of drinking and driving, speeding, not wearing seat belts, fatigue, and distracted driving (focusing on cell phone use and texting). Chapter Four also explores how some of these behaviors specifically affect motorcycle riders, since some behaviors are more dangerous when riding a two-wheeled vehicle than a four-wheeled one.

Researchers study the effects of driver behavior in two broad ways: laboratory studies that assess driver performance in a simulated environment, and epidemiological studies that attempt to identify the causes of real-world crashes. Both have shortcomings. Simulation studies may not accurately reflect the real world, while studies based on crash results may not be

---

[3] We are not aware of any studies that attempted to conduct similar analyses on the civilian population, since crash databases do not contain information on drivers' previous behavior. As discussed later in this chapter, some factors that affect crashes are difficult to study because of lack of data.

able to analyze all the factors that contributed to them. This is a particular issue with studying fatigue and districted driving, which can be difficult to determine after a crash, especially a fatal one.

Two caveats are important when interpreting the risk of any particular behavior. First, since many factors influence the occurrence and severity of crashes, it is difficult to assign specific weights to each. In most cases, researchers try to isolate one factor and determine its impact while holding other factors constant, but in the real world many factors come into play simultaneously. In particular, young men are more likely to engage in most of these risky behaviors than older men or women, and this helps explain their far higher crash rates (Vanderbilt, 2008). Second, in driving as in other areas, risky behaviors tend to occur in combinations. Drivers who drink, for example, may also be less likely to wear seat belts, and in the event of a fatal crash both behaviors would be considered contributing factors (Subramanian, 2005; Nichols, Chaudhary, and Tison, 2009). This discussion represents the general consensus of the traffic safety research community on how risky behavior affects crash occurrence and severity.

Drinking and Driving

It is well-established that drinking impairs driving. As one review put it, "There is clear and abundant evidence linking alcohol-involved driving to the increased likelihood of crash occurrence and severity," dating back to the 1960s (Sivak et al., 2006, p. 17). Alcohol has both physical effects, such as slowing reaction time, and psychological effects, such as increasing aggression and impulsiveness, that can also hinder safe driving.

The United States has been successful in the past in decreasing the rate of drunk driving. In particular, rates of drunk driving among people under 21 fell in the 1980s at an even greater rate than a general decline in drunk driving. This decline is attributed to four factors: a shift in the age distribution of the population, an increase in the minimum drinking age to 21, lower blood alcohol concentration (BAC) levels for classifying younger people as drunk (0.02 as opposed to 0.08), and general anti–drunk driving campaigns (Hedlund, Ulmer, and Preusser, 2001).

Despite these efforts, driving while under the influence of alcohol continues to be a major cause of fatal vehicle crashes. The absolute number of alcohol-related fatalities fell during the 1980s, from about 20,000 per year to between 13,000 and 15,000, and it has remained roughly constant since the early 1990s. Since 1998, the percentage of crash deaths attributed to alcohol has remained steady at about 30 percent (NCSA, 2009a).

Young adults have the highest rates of drunk driving and alcohol-related crashes of any age group (Sivak et al., 2006). In 2008, for example, of drivers aged 21–24 who were involved in fatal crashes, 34 percent had BAC levels over 0.08,[4] which is the current legal limit in all states. The average for all drivers was 22 percent; for teenagers, it was 17 percent (NCSA, 2009a). The average for men of all ages was 25 percent, and for women, 13 percent.

Some research has suggested that the same BAC level can be more dangerous in younger than older drivers. One study found that the crash risk of men 16–20 begins rising steeply at much lower BAC levels than it does for women that age or for older drivers. By BAC 0.08, the

---

[4] BAC is typically measured in grams per decaliter (g/dL). For the sake of simplicity, throughout the paper we will drop the g/dL and simply discuss BAC as a number.

odds ratio[5] of being in a fatal crash is 10 for most drivers, compared to sober drivers the same age; for 16- to 20-year-old men, the odds ratio is about 30 (Zador, Krawchuk, and Voas, 2000). Peck et al., 2008, looked at age and gender separately; they found that drivers under 21 have a greater risk for crashes at any BAC level over 0, while drivers over 21 do not have an increased crash risk until their BAC levels reach about 0.06. Figure 3.1 shows these relative odds ratios of crashes for age and gender groups for both studies.

Speeding

Speeding affects crashes in two ways: it increases the overall crash risk and it increases the severity of crashes. The overall crash risk increases because a driver who is speeding has limited ability to make an evasive maneuver and a shorter stopping distance. Crash severity increases as a law of physics; crashes at higher speeds involve greater energy. Almost one-third of all fatalities in U.S. crashes are caused by speeding (Sivak et al., 2006).

Research has conclusively found that crash risk increases with each increment of speed over the average traffic speed. This relationship is exponential rather than linear; that is, the risks increase only a little at speeds just above average, but increase more and more quickly at higher speeds. As a rule of thumb, a "1% increase in speed results approximately in 2% change

Figure 3.1
Relative Crash Risks by Blood Alcohol Concentration Level by Age and Gender

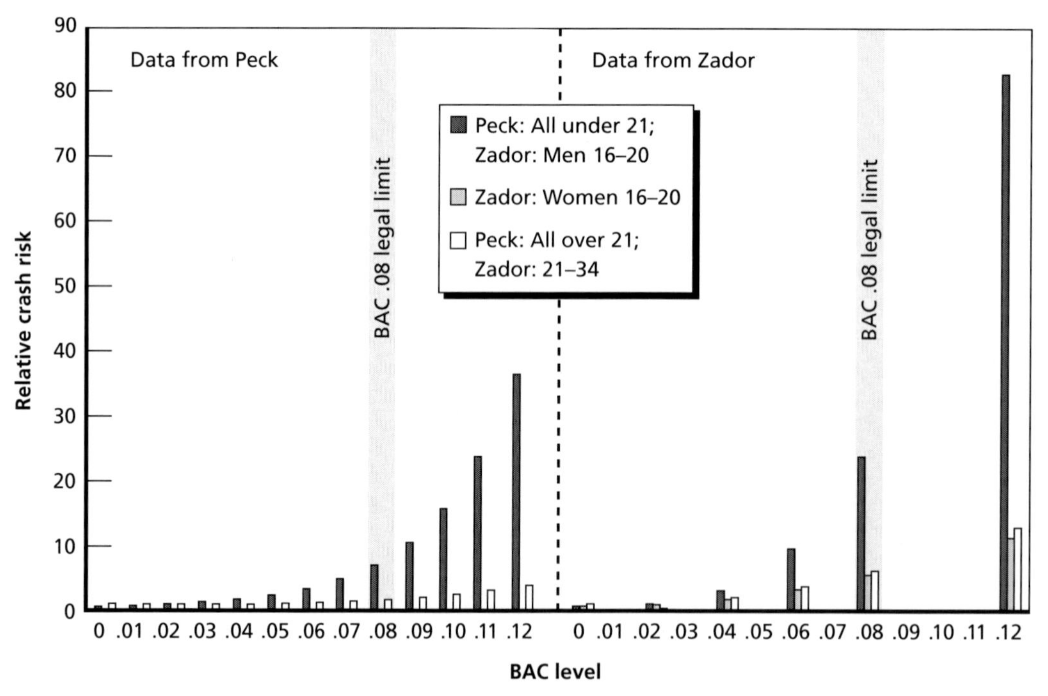

SOURCES: Zador, Krawchuk, and Voas, 2000, Table 6.6; Peck et al., 2008.
RAND *TR820-3.1*

---

5 An odds ratio provides a rough sense of how different a condition varies from a norm, which is designated as 1. An odds ratio of 1.5 means the condition carries approximately a 50 percent increase that a given event will occur. See the section below "Factors Identified in Studies of Military Personnel" for a discussion of interpreting odds ratios.

in injury crash rate, 3% change in severe crash rate,[6] and 4% change in fatal crash rate" (Aarts and van Schagen, 2006, p. 223, citing work by Nilsson, 2004). The actual rate of increase depends on the nature of the road; crash risks increase faster on urban than rural roads and on minor roads than major roads (Aarts and van Schagen, 2006).

The severity of crashes also increases with speed. As Figure 3.2 shows, as change in speed at collision rises,[7] the probability of fatality as well as the number of injuries per 100 vehicle occupants rises. While these two sets of figures (the probably of fatality and the number of injuries per 100 occupants) are from different studies and not directly comparable, showing them in the same figure illustrates how all indicators of crash severity rise with speed. Fatal injuries do not rise as quickly as less serious ones at lower speeds, but at higher speeds they

Figure 3.2
Risks of Moderate and Serious Injury per 100 Occupants and Probability of Fatality Based on Change in Speed at Impact

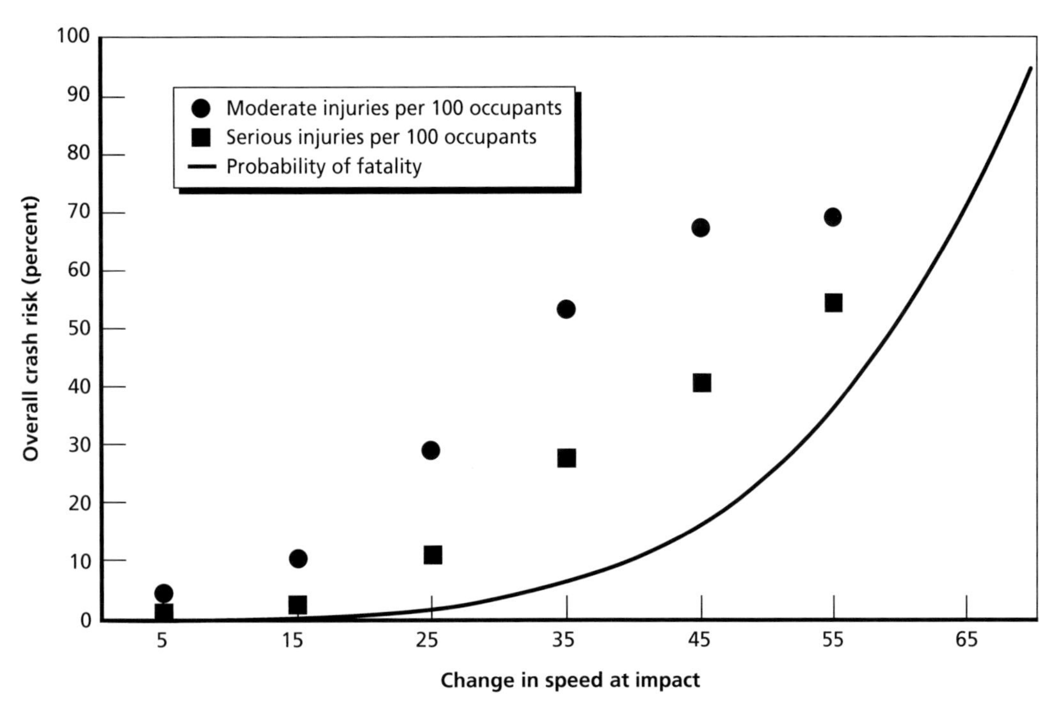

SOURCES: Stuster and Coffman, 1998, based on data from Bowie and Waltz, 1994, on moderate and serious injury risk and Joksch, 1993 on fatality risk.
NOTE: The data points for the moderate and serious injuries are based on categories; the data point at 5 represents the category 0–10 mph, and so forth; 55 represents speeds 50 mph or more. "Moderate" includes any injury at 2 or above on the six-level abbreviated injury scale (where 1 is minor and 6 is fatal); "serious" is a 3 or above.
RAND *TR820-3.2*

[6] While these terms were not defined in the paper cited, generally an injury crash is one that produces any type of injury, even a minor one, whereas a severe crash produces a serious injury (perhaps requiring hospitalization or producing disability) or substantial property damage.

[7] Figure 3.2 is based on change in vehicle speed at the point of impact, which is different than the speed before impact. The change in speed refers to the difference between the vehicle's pre-impact speed and its speed immediately following impact. If a vehicle traveling 55 mph crashes into another moving vehicle, and its speed post-crash drops to 20 mph, its change in speed is 35 mph.

increase dramatically. Fatalities begin to increase with a change in speed of 30 mph, and by 60 mph half of all crashes cause fatalities. The chance of dying in a crash is 15 times higher at 50 mph than at 25 mph (Stuster and Coffman, 1998).

Researchers agree about the risks of speeding, but there is less agreement as to whether driving more slowly than average results in increased crash risk. Some research has found that vehicles moving much more slowly than surrounding traffic are at greater crash risk, while others have not. One theory is that the risk associated with driving more slowly is related to turning and other maneuvers that require slowing down to execute safely. This theory is supported by the fact that only 5 percent of fatal crashes happen between two vehicles traveling in the same direction, and far more often between vehicles going in different directions (Stuster and Coffman, 1998).

One cause of driving too slowly is being drunk; in one study, almost half of drivers pulled over for driving 10 mph or more under the speed limit were over an 0.08 BAC level (Stuster, 1997). However, speeding is not necessarily associated with drinking; in that same study, only 9 percent of those driving more than 10 mph over the limit were over 0.08 BAC (Stuster, 1997). In another study, only 10 percent of motorcyclists speeding were over 0.08 BAC (Stuster, 1993).

Young men are more apt to be involved in speed-related crashes than other demographic groups. One study found that 40 percent of crashes caused by male drivers 15 to 20 years old were the result of speeding. Speeding as a factor contributing to crashes declines with age (Stuster and Coffman, 1998).

Seat Belt Use

Over half of drivers killed in crashes are not wearing seat belts, making lack of seat belt use the single biggest contributor to fatal crashes. As one report put it, "The single most effective technology for reducing injury severity in a motor-vehicle crash is the safety belt" (Sivak et al., 2006). One estimate found that wearing a three-point seat belt reduces the risk of fatality by 45 to 60 percent (Kahane, 2000), meaning that for every 100 people killed who were not wearing a seat belt, between 45 and 60 would have survived if they did. NHTSA estimated in 2006 that wearing seat belts had saved over 15,000 lives in that year alone, and that more than 5,000 people who died in crashes could have survived had they been wearing seat belts. In contrast, air bags reduce the risk of death by 14 percent without a seat belt and by 11 percent with a seat belt (NCSA, 2006).

Seat belt use in the United States has risen from 73 percent in 2001 to 84 percent in 2008, although the rate varies considerably from state to state (NCSA, 2009b). This is still below usage rates in comparable countries, such as Australia and the United Kingdom, where use is well over 90 percent (Sivak et al., 2006). Seat belt use varies to some extent with demographic variables; see Table 3.1.

Fatigued Driving

Fatigued[8] and distracted driving are both considered forms of "driver inattention," meaning drivers who divert their attention, even momentarily, from the task of driving for any reason.

---

[8] *Fatigued, drowsy,* and *sleepy* driving are all used in the literature to mean driving while feeling tired or actually fighting off sleep. While some researchers define these terms more precisely, for our purposes the three are basically interchangeable, and we have chosen the term *fatigued driving.*

Table 3.1
Difference in Seat Belt Use by Selected Factors, 2008

Factor	% Seat Belt Use
Gender	
Men	81
Women	86
Age	
16 to 24	80
Over 25	84
With or Without Passengers	
Solo drivers	82
Drivers with passengers	87
Drivers 16 to 24	
With passengers the same age	80
With passengers younger or older than themselves	87
Race	
Black	75
White	83

SOURCE: NCSA, 2009a.

Driver inattention is more difficult to study than drunk driving or speeding, for three main reasons. First, no definitive test exists to assess inattention among drivers. Determining whether a driver who has been involved in a crash was fatigued or distracted is difficult; there is no equivalent to a breath analysis test that determines BAC levels. Second, researchers speculate that police are reluctant to attribute crashes to driver inattention without proof or an admission of guilt from drivers, who are often reluctant to admit that they were distracted or fatigued. Finally, instances of fatigued driving and distracted driving are not compiled in any national database such as FARS, making it hard to analyze its prevalence and effects.

As a result, estimates differ greatly on the percentage of crashes caused by driver fatigue. One study, based on an analysis of the NHTSA Crashworthiness Data System, found that fatigue contributed to about 4 percent of all crashes. However, information on whether the driver was considered inattentive was missing on many of the data forms, so it is possible that rates are higher (Stutts et al., 2005b). Others have put the percentage of crashes attributable to fatigue much higher, at approximately 20 percent of crashes (Smith et al., 2009).

Some common elements that investigators use to determine whether fatigue was involved include the time of day (most crashes due to fatigue occur late at night, in the early morning, or the mid-afternoon), whether the vehicle left the roadway, and whether the driver attempted to avoid the crash (an alert driver braking to avoid a crash would leave skid marks, while a driver who has fallen asleep would not). Crashes due to fatigue also more often involve solo drivers, without passengers (Strohl et al., 1998). Laboratory studies have found that drivers do not always accurately assess their own level of fatigue (Strohl et al., 1998).

Although estimates of the degree of risk vary, there is clear evidence that driving while fatigued is significantly more dangerous than driving while alert. An NHTSA study[9] took real-time observations from 100 vehicles that were driven under real-world conditions over the course of a year to study in detail the causes of crashes and near crashes (events in which the driver's actions prevented a crash from occurring). Klauer et al. (2006, pp. 28–30) found that fatigued drivers were more than six times more likely to be involved in a crash or near crash than drivers who were not fatigued. The only activity with a higher likelihood of causing a crash was reaching for a moving object. Ingre et al. (2006) found that drivers who scored 8 on the Karolinska Sleepiness Scale (where 1 is fully alert and 9 is fighting sleep) were 28 times more likely to get into a crash than alert drivers, while drivers measuring 9 were 185 times more likely.

Driving while fatigued presents additional risks beyond driving while distracted. First, while many drivers can be distracted and still drive without incident, drivers who become fatigued generally have problems driving. Klauer et al. (2006, p. 24) found that of all the time drivers spent behind the wheel without incident, they were fatigued only 2 percent of the time. However, for crashes and near crashes, drivers were fatigued about 10–14 percent of the time. For other types of inattention, such as eating or adjusting the radio, the percentages of time were more similar, meaning that drivers were engaged in distracting activities without getting into a crash. Second, crashes caused by fatigue tend to be more severe than those caused by other factors, possibly because they tend to occur at higher speeds (Strohl et al., 1998).

Crashes caused by fatigue are higher among younger drivers than those in their mid-20s and older, with a greater percentage among men than women. Theories to explain this include a greater need for sleep, changes in sleep patterns, and cultural and lifestyle factors (Strohl et al., 1998). One study suggested that the relative inexperience of younger drivers and, in particular, their lack of experience in identifying hazards while driving make them more vulnerable. A performance study found that younger drivers' ability to recognize hazards when they were only mildly fatigued was impaired, whereas for more-experienced drivers, mild fatigue did not affect this ability (Smith et al., 2009). A similar study found that while younger drivers had faster reaction times than older drivers while alert, their reaction times slowed when they were fatigued while older drivers' reaction times were not affected (Philip et al., 2004).

Although most fatigued driving is due to insufficient sleep or sleep disorders, alcohol can exacerbate the effects of fatigue. Studies of crashes by fatigued drivers found rates of alcohol use (not necessarily at the legal limit) ranging from 20 percent (Wang, Knipling, and Goodman, 1996, looking at single-vehicle crashes only) to about one-third (McCartt et al., 1996). One simulator study compared driving with four or eight hours of sleep and no alcohol to driving with four or eight hours of sleep and a small amount of alcohol. Drivers who were sober all performed well behind the wheel, but drivers who slept eight hours and had a drink were four times more likely to leave the road, and those who slept only four hours and had a drink were 15 times more likely to leave the road (Roehrs et al., 1994).

---

[9] This project is called the 100 Car Naturalistic Driving Study. Because of the size of the project and the large amount of data collected, multiple reports have been published with the data from the study. For this report, we reviewed Klauer et al. (2006) and Dingus et al. (2006).

Distracted Driving

Distracted driving is a growing safety concern, given the prevalence of electronic devices that can be used by drivers. Not every distraction diverts drivers' attention to such a degree as to be unsafe, but studies of distracted driving include those activities thought most likely to impair driving, with electronic devices being a chief concern.

As with fatigued driving, estimates vary with regard to the percentage of crashes caused by distracted drivers. Information on whether drivers were using electronic devices is not generally supplied in the police reports that serve as the foundation for databases on crashes. According to one recent estimate, based on analysis of the NHTSA Crashworthiness Data System, 12 percent of all crashes were caused by distracted drivers, with an additional 10 percent involving drivers who "looked but didn't see," (the study did not specifically identify cell phone use as a cause of crashes). Information on whether the driver was considered inattentive was missing on many of the data forms. Therefore, these can be considered minimum estimates (Stutts et al., 2005a). The NHTSA study on real-time observations from 100 vehicles found that use of an electronic device contributed to 8 percent of the crashes and 5 percent of the near crashes (Dingus et al., 2006). NHTSA found that the percentage of fatal crashes in which drivers were distracted rose from 11 to 16 percent from 2004 to 2008. Over the same period, injury crashes due to distraction declined from 26 to 22 percent (NCSA, 2009c). The National Safety Council developed a model of the percentage of crashes based on NHTSA estimates of the percentage of drivers assumed to be using their cell phone while driving and estimates from the literature on relative crash rates. The council estimated that cell phone use was responsible for 25 percent of all crashes, and texting for another 3 percent (National Safety Council, 2010).

These estimates presumably vary widely in part because cell phone use and texting continue to grow, but also in part because they are sensitive to assumptions. The National Safety Council estimate assumes, for example, that the increased risk of being in a crash while talking on a cell phone is four times higher than not talking on a phone, but estimates of the danger posed by cell phone use are quite wide, as noted below. In addition, the National Safety Council model uses the NHTSA figure that 11 percent of drivers are on the phone at any given moment, but this figure is specifically for daytime use (NCSA, 2009c) The National Safety Council model assumes this figure holds true for all driving, but this may or may not be true. Without better data on this important topic, it is difficult to estimate the true risk or the extent of the problem.

While the amount of cell phone and text message use has increased dramatically over the past decade,[10] over the same period the absolute number of deaths has declined, especially in the past several years. This seems counterintuitive: If a risky behavior has been gaining in popularity, we might expect overall deaths to increase, but the opposite has happened. It is possible that the decline in driving risk is due in large part to the decrease in VMT over the past several years, but we did not find any research looking at the relationship between crash risk, VMT, and distracted driving. Other factors that contribute to declining traffic deaths, such as vehicle safety features, have probably not changed as quickly over this same time period.

---

[10] According to CTIA, a trade association for the wireless industry, the number of annualized cell phone minutes grew from 259 billion in 2000 to 2.3 trillion in 2009. The number of text messages sent increased from 81 billion in 2005 to 1.6 trillion in 2009 (figures for 2000 were not available). See CTIA the Wireless Association, 2010.

Scores of studies have been conducted recently on the effects of driver distraction. Most of these are about cell phone use; we identified only a handful of studies dealing specifically with texting, probably because the phenomenon is much newer.

Cell Phone Use. The studies on cell phone use are clear on two points: Using a cell phone (talking and dialing) increases reaction time, and handheld and hands-free phones carry similar risks. There is less agreement on exactly how dangerous cell phones are on the road because such usage is more difficult to study.

Talking on a cell phone while driving is dangerous chiefly because it increases reaction time. One meta-analysis of 28 studies found that talking on a cell phone increases reaction time by over one-tenth of a second (Horrey and Wickens, 2006). Caird et al. (2004) found in a meta-analysis of 22 studies that the average reaction time of drivers using a cell phone increased by almost one-quarter of a second on average, and by a half-second for older drivers. Given that a car driven at 60 miles per hour covers about 90 feet per second, every one-tenth of a second increase in reaction time means that the car will travel an additional 9 feet before the driver brakes. Talking on a cell phone does not have a major impact on driving speed (Caird et al., 2004) or on staying in a lane (Horrey and Wickens, 2006).

Most studies found no difference between handheld and hands-free cell phones, suggesting that the distraction stems from the conversation itself, rather than the act of dialing or answering the phone (Caird et al., 2004; Horrey and Wickens, 2006). There has been limited research and mixed results about the effect of banning handheld and hands-free cell phones, as it seems many of the bans are not effectively enforced (McCartt, 2009). A review of studies of handheld cell phone bans in New York, the District of Columbia, Connecticut, and California, which compared insurance claims for crashes with those in neighboring states for one to two years after the ban was enacted, found that the bans had no effect on crashes (Highway Loss Data Institute, 2009).

However, the evidence is less clear on whether cell phones present a greater distraction than conversations with passengers. Some research has found that conversations with passengers distract drivers as much as cell phone conversations (Horrey and Wickens, 2006). Other research has found that passenger conversations are less distracting, since passengers tend to moderate their conversation based on driving conditions, while people on the other end of a cell phone are not in the same environment as the driver (Drews et al., 2008). In the NHTSA study, Klauer et al. (2006) found that passengers actually reduced the risk of being in a crash by about half; this study did not compare handheld to hands-free devices.

In epidemiological studies, driver cell phone use has been linked to higher crash rates. However, it is difficult to prove whether distraction actually causes these crashes, and it is not clear how much drivers' risk increases when they are on the phone. One review found estimated increases in risk for a crash from use ranging from 16 to 900 percent (Sugano, 2005). Klauer et al. (2006), using real-world data from the NHTSA driving study, found that *talking* on a cell phone did not statistically increase the risk of being in a crash. *Dialing* the phone more than doubled the risk. However, since drivers spent far more time talking on cell phones than dialing, both of these activities were found to account for about 3.5 percent of all crashes and near crashes.

There are two reasons that this topic is difficult to study. First, most jurisdictions do not require police reports of crashes to record the use of cell phones by the parties involved in the crash, so many databases do not yield sufficient information for study (Caird et al., 2004). Second, many studies do not control for amount of cell phone use or demographic variables.

Laberge-Nadeau et al. (2003) found that among drivers who use cell phones, the risk of being involved in a crash is 38 percent higher than for drivers who do not. However, when they controlled for other factors, the risk decreased to 10 percent higher for men and 20 percent higher for women.

With regard to the amount of use, it is not clear whether younger drivers are more distracted by cell phones than older drivers, or whether they simply use cell phones more often. The review by Young and Regan (2007) suggests that drivers in their teens and 20s are less experienced than drivers over 30 and thus are more susceptible to distraction. In Caird and colleagues' (2004) review of performance studies, age did not affect how well drivers performed while on a cell phone. In their real-world study of drivers, Dingus et al. (2006, p. 167) found that inattention-related crashes were up to four times higher among drivers age 18 to 20 than in drivers over 35. However, this includes inattention of all types (not only electronic devices) and does not control for the possible effects of higher cell phone use in that age group.

Texting. In comparison to the large body of research on using cell phones, few studies have been done specifically of sending and receiving text messages. Three studies that looked only at the effects of texting on young drivers (under 24) all found that the primary problems caused by texting were staying in the lane, reaction time, and not keeping a steady distance from the vehicle in front of them. Two found that sending texts impaired driving ability more than receiving texts, while the third found the reverse.

Some specific findings from each study:

- Drivers were 45 to 70 percent more likely to leave their lane when sending texts than when not distracted, but receiving texts had no effect. The amount of time drivers were not looking at the road increased from 10 percent while not distracted to 40 percent while texting (Hosking et al., 2006).
- Reaction time while sending a text increased by about 35 percent; for one experiment, the average reaction time increased from 1.2 to 1.6 seconds, which would increase stopping distance by three car lengths. These increases were greater than those for drinking and driving and hands-free conversations, but less than for handheld phones. Finally, drivers strayed from their lane ten times more often when texting (Reed and Robbins, 2008).
- Reaction times increased by 30 percent while texting, and the rate of being involved in a crash increased six-fold while texting (Drews et al., 2009).

Factors Identified in Studies of Military Personnel

While there has been far less analysis of motor vehicle crash fatalities among military personnel than among the overall population, several studies have specifically looked at military personnel. This section relies heavily on five recent studies, all of which looked at somewhat different aspects of the problem:

- Bell et al. (2000b) reviewed motor vehicle crashes among Army personnel that led to hospitalization (on the grounds that these represented serious injury) from 1992 to 1997.
- Hooper et al. (2006) looked at motor vehicle fatalities for men in all services from 1991 to 1995.

- Wilson et al. (2003) examined all types of accidental deaths sustained by Army men from 1990 to 1998.
- Carr (2001) looked at fatal and severe injury crashes (those that caused permanent disability) among Air Force personnel from 1988 to 1999.
- Bowes and Hiatt (2008) studied vehicle crash deaths among enlisted Marines, separating results by motorcycle crashes and all other vehicle-related deaths, from 1998 to 2007.

While Wilson et al. (2003) did not separate motor vehicle fatalities from those resulting from other types of accidents, some of the risk factors they identified among the best predictors of accidental death were clearly related to driving behaviors. In addition, about two-thirds of accidental deaths in the military are motor vehicle fatalities (Powell et al., 2000).

Other than Bowes and Hiatt (2008), we did not locate any other studies of motor vehicle crash rates or fatalities since Operation Enduring Freedom/Operation Iraqi Freedom (OEF/OIF) started.[11] Several OEF/OIF–era studies have looked at the prevalence of risky behavior, which may affect crashes; these are discussed in Chapter Five.

Four of these papers compared deaths or injuries to other personal information—such as personnel records or responses to the Health Behavior Survey—to help "predict" which characteristics put military personnel most at risk. They used these to develop "adjusted odds ratios," meaning that they compared two or more groups with one serving as the base case, and adjusted them to exclude the effects of other variables, such as age or gender. The odds ratios are expressed as a number; the closer that number is to one, the less difference between the reference group and the other group(s). While this should not be interpreted as the literal odds, the further the number is from one, the stronger the relationship is between the two variables.[12] Hooper et al. (2006) developed odds ratios based on deployment status[13] and enlistment, not for the entire group; these will be called out in the discussion below. In all cases, the researchers developed these measures while controlling for other variables.

Other types of analyses included Bowes and Hiatt (2008), who constructed a set of mortality figures by age group. Carr (2001) did not link crashes to other data, but instead analyzed how the causes of crashes for Air Force personnel compared with those in the general U.S. population.

Table 3.2 presents these odds ratios for the various factors in descending order based on how strongly they are associated with vehicle crashes and/or deaths. The higher the factor in the list, the stronger the evidence that it predicts who will be in a vehicle crash.

---

[11] The Veterans Administration is planning a study of this type, but as of this writing research had not yet begun (Aaron Schneiderman, email to Liisa Ecola on January 19, 2010).

[12] We also looked at confidence intervals, which indicate the likelihood that the result is not the result of chance. Unless otherwise indicated, all of the results reported here have confidence intervals that did not include zero, meaning that even if the effects are modest, they are statistically significant.

[13] A number of studies looked at the impacts of deployment, defined as deployment of a service member to a combat zone at least once during a conflict. We did not locate any studies that looked at the effects of one versus multiple deployments. Fear et al. (2008) uses factors such as length of deployment and time of deployment (whether in a conflict or post-conflict phase) as explanatory variables; this study is discussed in Chapter Five.

Table 3.2
Odds Ratios for Vehicle Crashes Found for Demographic and Behavioral Factors Among U.S. Military Personnel

Factor	Odds Ratios in Literature	Comments
Found to contribute to higher crash rates (in descending order)		
Age (generally under 25)	4.9[a], up to 2.7[b]	Other studies also found strong effect but did not report an odds ratio; age categories varied by study
Enlistment with a felony waiver	4[d]	Motorcyclists only
Being a motorcycle rider	2.2[c]	Finding held true for all types of accidental death, not just vehicles
Prior hospitalization for substance abuse	2[b]	Only for nondeployed
Drinking heavily	1.8[a], 1.7[c]	
Not using a seat belt regularly	1.4[a], 1.3[c]	
Getting less than six hours of sleep regularly	1.3[c]	
Deployment	1.3[e], 1.8 to 1.3[d]	For OEF/OIF deployments, 1.8 3–6 months after return, 1.3 6–9 months after return
Occupation	1.4[c], up to 2.5[d]	Higher for combat occupations[c]; higher for infantry and human resources/admin/finance but only for motorcycle crashes[d]
Lower educational attainment	1.2[c], up to 2.2[b]	High school diploma or less compared to some college; about half the risk with some college compared to high school diploma, but no high school diploma also had lower odds[d]
Marital status (single or divorced vs. married)	1.3[c]	"Honeymoon effect" with lower odds (0.53) for those married less than one year, for car drivers only[d]; marital status made a difference only for those who deployed[b]
Recently promoted or new command	0.6[d], 0.4[d]	
Passing scores on physical fitness exams	As low as 0.3[d]	For car drivers only
Mixed evidence or did not have a significant effect		
Race		1.8 for nonwhites[a]; lower rates for blacks and motorcycle deaths[d]; others found no difference[c]
Gender		Gender made no difference[a]; higher rates for 19–20 year old women only[d]
Service branch		No difference after adjusting for demographics[b]; no difference between active duty and reserve/guard[b]
Officer or enlisted		2.5 odds ratio for enlisted compared with officers among deployed; no difference for nondeployed[b]; no difference between enlisted, warrant officers, and officers[a]; higher odds for warrant officers and E1/E2 Marines six months after enlisting[d]
Driving while drunk		Conflicting findings: no increase in risk for those saying they have driven drunk[a]; or 1.4 odds ratio[c]

Table 3.2—Continued

Factor	Odds Ratios in Literature	Comments
Speeding		Conflicting findings: no increase in risk for those saying they routinely go 11 or more miles per hour above the speed limit[a], or 1.4 odds ratio[c]
Other behavioral factors		Number of miles driven, previous hospitalization for vehicle crashes, being under stress or dissatisfied with life, or having a support network of friends had no effect[b]

SOURCES: (a) Bell et al., 2000b, motor vehicle crashes that led to hospitalization in the Army, First Gulf War; (b) Hooper et al., 2006, motor vehicle fatalities for men in all services (note that this study developed multiple odds ratios for all veterans, enlisted only, and deployed versus nondeployed, so we have reported in some cases the highest ratio from among those four), First Gulf War; (c) Wilson et al., 2003, all types of accidental deaths sustained by Army men, First Gulf War; (d) Bowes and Hiatt, 2008, Marine deaths in motor vehicle crashes, OEF/OIF; (e) Knapik et al., 2009.

NOTE: All odds ratios have been rounded to one decimal place.

Demographic Factors

Age is the most significant predictor of motor-vehicle crash deaths. As in civilian populations, younger drivers are several times more likely to be killed in crashes than older ones. Bell et al. (2000b) found a 4.9 odds ratio for soldiers under 21 (compared to those over 40). Hooper et al. (2006) found higher odds for all groups (compared to those under 25); the highest odds ratio was nondeployed enlisted at 2.7 (compared to those over 36). The odds ratios decreased with increasing age. Bowes and Hiatt (2008) did not develop odds ratios, but they presented a chart showing the number of Marines killed in vehicle crashes per 100,000 persons (similar to Figure 2.4 in Chapter Two of this report). The mortality rate was highest for 19-year-olds (approximately 43 persons per 100,000), declined for those in their 20s, and stabilized at around age 30 in a pattern very similar to that of civilian men.

Results for gender and race were more varied than civilian statistics. Wilson et al. (2003) and Hooper et al. (2006) studied only male populations and found no relationship between race and death rates. However, Bell et al. (2000b) found a 1.8 odds ratio for nonwhites being hospitalized for crashes, as compared to whites. It is not clear why this would be the case; the researchers do not have an explanation except that there may be some unmeasured variable at work, such as occupation. A precursor to the Bowes and Hiatt (2008) study, Boning and Bowes (2003), found a 1.3 odds ratio among blacks and Hispanics, but Bowes and Hiatt found no such differences based on race or ethnicity.

With regard to gender, Bell et al. (2000b) found that sex made no difference in crash rates, and they suggest that women in the Army drive more than civilian women do. Bowes and Hiatt (2008) found that female Marines aged 19 and 20 are twice as likely to be killed in car crashes than women of other ages, whereas men were only 1.7 times more likely. However, given the low number of women who died in crashes overall (2.1 percent of all Marine crash fatalities), these results were not statistically significant.

Lower educational attainment and being single or divorced were generally associated with higher death rates than higher education and being married. Wilson et al. (2003) found, among all service members, an odds ratio of 1.2 for personnel with only a high school education as compared with those with at least some college. Hooper et al. (2006) found that for enlisted personnel, having only a high school education or less carried nearly double the odds than enlisted personnel with some college. Among enlisted Marines, having some college was associated with about half the level of risk as a Marine with a high school diploma or equivalent, but these results were not significant (Bowes and Hiatt, 2008). Wilson et al. (2003) and Bowes and Hiatt (2008) found increased odds for single over married personnel (1.3 and 1.5); Bowes and Hiatt also found a "honeymoon" effect in which Marines married less than one year had about half the risk of those married for other lengths of time. Hooper et al. (2006) found that for deployed personnel, being divorced, widowed, or separated carried an even greater risk than being single, whereas for those who did not deploy, this factor did not make a difference.

There was limited association with the branch of service and insignificant differences between enlisted personnel and officers. Only one study looked at all four branches of the services and found that while the Marines had higher odds ratios for crashes and the Air Force lower odds ratios overall, when the data were adjusted for other variables, the service branch did not have a significant impact (Hooper et al., 2006). Hooper et al. also found that for deployed personnel, the odds ratio for enlisted members as compared with officers was 2.5, but this factor did not make a difference when looking at service members who did not deploy.

Bell et al. (2000b) found that warrant officers and enlisted personnel did not have significantly higher odds ratios than officers. Bowes and Hiatt (2008) found fairly high odds ratios for warrant officers compared with senior enlisted personnel; close to 3 for car crashes and over 6 for motorcycle crashes, but these reflected low numbers of total fatalities and were not statistically significant. They also found that pay grades E-1 and E-2, the most junior enlisted service members, had a 3.6 odds ratio after six months or more in the Marines. They suggested that this may be a result of the limits on personal freedoms during basic training, which are lifted afterward.

In terms of occupations, Wilson et al. (2003) found that those in combat occupations had a 1.4 odds ratio for accidental death, as opposed to other support occupations. This was lower than the odds ratio Wilson et al. (2003) found for deployment, which was 1.7 (more will be included about the impacts of deployment in the next subsection). No study found differences between reserve and active duty components.

Bowes and Hiatt (2008) looked at several factors available from Marine personnel records. Those found to be statistically significant included low odds ratios (0.4 for both car and motorcycle crashes) within three months of reporting to a new command, lower odds for those promoted within the last three months (0.6 for car crashes and 0.2 for motorcycle crashes), and high odds (3.6) of death on a motorcycle within three to six months after demotion.

Behavioral Factors

In terms of behavioral factors, the most significant factor in accidental death was riding a motorcycle. Wilson et al. (2003) found an odds ratio of 2.2 over nonriders, even when controlling for age, race, marital status, education, occupation, deployment, and previous hospitalization. Significantly higher odds ratios were also found with drinking heavily (1.8 in Bell et al., 2000b, and 1.7 in Wilson et al., 2003), not using a seat belt regularly (1.4 in Bell et al., 2000b, and 1.3 in Wilson et al., 2003), and getting less than six hours of sleep per night (1.3 in Wilson et al., 2003). Perhaps surprisingly, Bell et al. (2000b) found essentially no increased risk with drivers who routinely drove 11 or more miles per hour above the limit, or with those who drove drunk. Wilson et al. (2003) found a 1.4 odds ratio for both. Hooper et al. (2006) found odds ratios above 2 for prior hospitalization for substance abuse, but this factor was only significant for nondeployed personnel. No other behavioral factors—number of miles driven, previous hospitalization for vehicle crashes, being under stress or dissatisfied with life, or having a support network of friends—had significant effects.

Bowes and Hiatt (2008) looked at several other factors not reviewed in the other studies. They found that having enlisted with a felony waiver increased the odds ratio of being in a motorcycle crash to 4, but other types of waivers (traffic, nontraffic, and drug and alcohol) did not show a significant impact on crashes. They also found that Marines who passed physical fitness exams were about half as likely to be in car crashes as those who do not have a passing score, but this had no impact on motorcycle deaths. The authors speculated that higher scores on physical fitness exams might indicate high personal discipline. Finally, they identified certain bases as having higher risks. For car crashes, the risks are highest at

- Camp Lejeune, N.C. (data also include Cherry Point/New River)
- Camp Pendleton in San Diego, Calif.
- The bases at Twentynine Palms, Barstow, and Yuma, Calif.
- The recruiting center at Parris Island, S.C.

For motorcycles, only Camp Pendleton was associated with more deaths. Deaths from car crashes are highest from July to December, while deaths from motorcycle crashes rise from April to September.

Carr (2001) looked at the causes of crashes that killed or disabled Air Force personnel and compared them with those causes among the U.S. population. Overall, Air Force crash rates were lower; if Air Force personnel had been in crashes at similar rates to those in the U.S. population, 600 more Air Force members would have died in the 12-year period. The difference is due largely to greater seat belt use among Air Force personnel, which was measured at 95 percent. The main causes of all crashes were drinking and driving (40 percent, although in some two-vehicle crashes the non–Air Force driver was drunk), speeding (39 percent), not using a seat belt (34 percent), and fatigue (19 percent). For motorcycle crashes, the main causes were speeding (48 percent), drinking (32 percent), not wearing a helmet (19 percent), and inexperience (16 percent). One major difference was that crashes in the general population were more evenly distributed throughout the week, while 46 percent of Air Force crashes took place between noon on Friday and noon on Sunday.

We identified one additional study specifically dealing with military personnel and fatigue, which looked at 58 fatal crashes among Finnish military conscripts from 1991 to 2004. As in the U.S. armed forces, vehicle crashes are the leading cause of death among conscripts in Finland. The study concluded that about one-third of the crashes were attributable to fatigue. This was the most prevalent cause; others included speeding (26 percent), drunk driving (23 percent), and suicides (18 percent). About two-thirds of the fatigue crashes occurred between midnight and 6 a.m. The cause of the crashes had previously been determined by multidisciplinary teams that investigate all fatal crashes in Finland, and the article did not describe their methodology. In a survey of conscripts, just over half reported having driven while fatigued during the previous two months (Radun and Radun, 2009).

Deployment

Deployment increased the risk of vehicle crashes. Numerous studies have been done of veterans from the Vietnam and first Gulf Wars; one study of OEF/OIF veterans confirmed this trend. For Gulf War veterans who deployed, there were 23.6 deaths from vehicle crashes per 100,000;

for those who did not deploy, the rate was 15.9 (Lincoln et al., 2006). A study by Knapik et al. (2009) reviewed 20 previous studies of accidental deaths that compared service members who deployed with those who did not. Sixteen of these studies were of American service members; two were from the United Kingdom and two from Australia. Their meta-analysis of these 20 studies found odds ratios of 1.26 for deployed veterans over those who did not deploy (the same odds ratio was found in both conflicts). While the meta-analysis included other forms of accidental deaths, the researchers noted that, "Much of the excess mortality among conflict-zone veterans was associated with motor vehicle events" (p. 231). None of these studies looked at the impact of multiple deployments.

Two findings from the Knapik review are of interest. First, in both the Vietnam and first Gulf wars, those studies that calculated increased risks separately for male and female deployed veterans found that women's odds of dying in a vehicle crash increased more with deployment than men's did. In Kang and Bullman (2001), deployed female Gulf War veterans had an odds ratio of 1.63 of being killed in a vehicle crash over nondeployed women, while for men the odds ratio was 1.19. Second, based on studies that followed service members up to 30 years post-deployment, the elevated risk of death seems to dissipate after five to seven years. In Boyle et al. (1987) and Boehmer et al. (2004), the odds ratio for dying in a vehicle crash was 1.93 within the first five years after deployment, but after six years, the difference was minimal. Knapik et al. suggested that accident mortality rates may decline over time because "risky behavior decreases as conflict-zone veterans 're-adapt' to civilian life" (2009, p. 19).

Bowes and Hiatt (2008) compared the effect of deployment on vehicle deaths for enlisted Marines before and after the launch of OIF.[14] In looking at all vehicle crashes (cars and motorcycles), the odds ratio for Marines in the period three to six months since deployment increased from 1.02 in FYs 1999–2002 to 1.8 in FYs 2003–2007. For six to nine months post-deployment, the odds ratio increased from 0.3 to 1.3. The statistical model did not track other time periods post-deployment, but Figure 3.3 shows the average number of crash deaths for the two periods.[15]

A number of theories have been put forward about why accidental deaths are higher among the deployed; none have been conclusively proven. First, there may be differences between those who deploy and those who do not (Bell et al., 2001); Chapter Five discusses some evidence that those who deploy may have taken more risks even before deployment and, therefore, would be at greater risk even if they did not deploy. Second, it may be due to post-traumatic stress disorder (PTSD) or other psychiatric conditions as a result of deployment. Third, those deployed may cope with stress by increased drinking or drug use, or they may have less regard for personal safety (Chapter Five reviews the evidence linking deployment and risk-taking). Fourth, there may be a reduced risk of surviving a crash because of previous illness or injury. The increased risk is specifically the result of accidents; those who deployed had lower death rates from illness than service members who did not deploy (Kang et al., 2002).

[14] To isolate the effects of OEF/OIF, they analyzed data from two periods: FYs 1999–2002 and FYs 2003–2007.

[15] As the total number of deaths is small, we also considered whether these odds ratios were statistically significant, meaning that the results were not simply coincidence. The 1.8 odds ratio was significant at 1 percent, meaning there is only a 1 percent chance that these results are a coincidence. While the 1.3 odds ratio is not itself statistically significant, the change from 0.03 to 1.3 is significant at 20 percent.

Figure 3.3
Number of Marine Deaths from Motor Vehicle Crashes, Before and After Major OIF Deployment

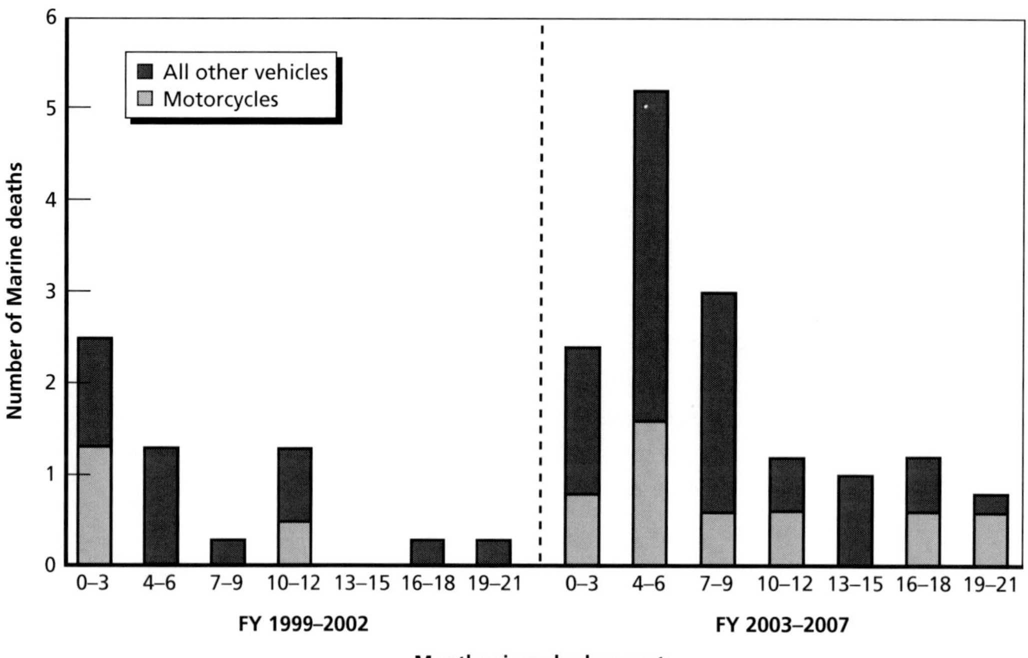

SOURCE: Bowes and Hiatt, 2008, Figure 13.
NOTE: Each bar represents the average number of deaths per year. For example, in the period FYs 2003–2007, on average 2.5 Marines who had been back from deployment less than three months died each year in vehicle crashes.
RAND *TR820-3.3*

Several other explanations have either been discredited or there has not been sufficient evidence to confirm or disprove them. One explanation—a hypothesis that nerve agent exposure during the first Gulf War may have led to neurological problems that resulted in worse driving—has been ruled out as the evidence did not support it (Gackstetter et al., 2006). Lincoln et al. (2006) did not find enough evidence to say whether some fatal crashes of Gulf War veterans were suicides. The same study suggested that fatigue could be a contributing factor to higher crash rates, since two of the indicators associated with fatigue-induced crashes—single-vehicle involvement and driving at night—occur more frequently among the crashes of deployed personnel than nondeployed personnel.

How Risks Differ Between Motorcycles and Cars

Bowes and Hiatt (2008) looked specifically at demographic characteristics describing military personnel involved in fatal motorcycle crashes as compared with those killed in passenger vehicle crashes.[16] While this study was limited to Marines, it found the interesting result that some factors that are key predictors of higher odds of dying in a car crash have no effect on motorcycle crashes and vice versa. For example, being recently married seemed to "protect" against dying in a car crash—that is, those odds ratios were lower than for single Marines. But recently

[16] This category also includes pedestrian deaths, but the number of such deaths is very small.

married men who rode motorcycles were just as likely to be killed in crashes as those who were single (that is, there was no "honeymoon effect.") Good physical fitness scores seemed to "protect" against dying in a car crash, but they had no impact on motorcyclists' odds. The study did not link health records with deaths, so it did not look at factors such as drinking, not wearing a seat belt, or other risk-taking behaviors.

This seems to suggest that those at higher risk for car crashes and those at risk for motorcycle crashes are separate groups. If we develop a composite portrait of a Marine at highest risk for being killed in a car crash, based on those factors most closely associated with car crash deaths, he or she is single with no dependents, fairly junior (an E1 or E2 six months or more after enlistment),[17] with low scores on the physical fitness exam, and who is more than three months past the most recent promotion or reporting to a new command and/or six to nine months post-deployment. In contrast, the Marine at highest risk for being killed in a motorcycle crash is a white man (in the ten years covered by this study, no women or blacks were killed on motorcycles) and an E5 or E6, working in personnel/administration, finance, or infantry/artillery/armor, who is more than three months past the most recent promotion or reporting to a new command, and who entered the Marines with a felony waiver. This may further suggest that campaigns and enforcement policies might be most effective if targeted to these specific groups.

[17] Pay grades for enlisted Marines range from E1, the most junior, to E9. This study reported risk by pay grade, not age; we assume that that E1 and E2 grades are generally in their late teens to early 20s, and E5 and E6 probably in their mid-20s to early 30s.

CHAPTER FOUR

Factors Influencing Motorcycle Crash Rates

Motorcycle riders have a high risk of crashing and being injured. NHTSA has estimated that 80 percent of motorcycle crashes injure or kill a motorcycle rider, while only 20 percent of passenger car crashes injure or kill an occupant (NHTSA, 2003). As this chapter discusses, there are three other broad differences from passenger cars as well. First, motorcyclists have some risk factors separate from car drivers. Second, many behaviors that are dangerous behind the wheel of a car are even more dangerous on a motorcycle. Third, riding a motorcycle is becoming more dangerous, while driving a car is becoming safer, and the risks are increasing, particularly for older riders.

It is not entirely clear from the literature why fatalities and fatality rates have increased so much during the past decade, a time when the number of car crashes has been stable or even decreasing. Several distinct trends seem to be occurring. Motorcycle ownership is increasing, and it is possible that new riders are less experienced. Fatalities among men over 40 have risen the most, perhaps implying that members of this group are taking more risks or riding more frequently. Sport and supersport bikes, which as discussed below are more dangerous than conventional bikes because of their light weight and high speeds, are gaining in popularity. Finally, helmet use declined in the first part of the 2000s, perhaps because of changes in helmet laws. However, no overall analysis has been able to state which, if any, of these trends is the largest contributor to these increases.

Many factors contribute to the risks of injury associated with motorcycle riding, only some of which can be modified. Factors that cannot be directly modified include the rider's age, gender and socioeconomic status; the type of bike and engine size; and the time of day, day of the week, and season of the year the motorcycle is being ridden (Lin and Kraus, 2009). However, information about these factors may be helpful in targeting prevention programs to the proper audience based on relevant age, gender, or socioeconomic status. Behavioral factors, which are more relevant when it comes to designing successful prevention programs, include wearing a helmet, the use of alcohol and other drugs, driver inexperience, conspicuity of the motorcycle and rider, driving without a license, ownership, speeding, and risk-taking behaviors (Lin and Kraus, 2009).

General Demographic Factors

As with cars, age and gender are strongly associated with fatal crashes for motorcyclists. But unlike cars, it is no longer younger motorcyclists that are at greatest risk for fatal crashes. In 2008, the most recent NHTSA data on fatal crashes, motorcyclists aged 40 and older were involved in far more fatal crashes than those under 30 (NHTSA, 2009a). Figure 4.1 shows the large increase in motorcycle fatalities for riders aged 40 and older over the past decade. From 1998 to 2008, fatal crashes for motorcyclists under age 30 increased by 75 percent, while for motorcyclists aged 40 and older fatal crashes increased by 254 percent (NHTSA, 2009a). The higher number of fatalities in older motorcyclists may be due in part to increased ownership and licensing of motorcyclists over the age of 40, which appears to be a worldwide phenomenon (NHTSA, 2003; Haworth and Mulvihill, 2005; Motorcycle Safety Foundation [MSF], 2005; Creaser et al., 2007). However, data on the number of fatalities per motorcyclist in each age group were not available, so it is difficult to say whether motorcyclists over 40 are at higher risk, or their numbers are simply growing.

It is not clear if the military is experiencing a similar trend toward older riders being at higher risk. A 2001 study of motor vehicle fatalities among U.S. Air Force personnel found that airmen under the age of 26 were "three times over represented for motorcycle fatalities compared to those 26 years of age and older" (Carr, 2001, p. 25). But Bowes and Hiatt (2008) found that motorcycle fatality rates were highest for Marines aged 27 and 32.

Motorcycle fatalities are overwhelmingly male, because motorcycle riders are mostly male. From 1995 to 2005, about 90 percent of all motorcycle riders killed were male (Shankar and Varghese, 2006). Even though the absolute number of female riders killed more than doubled during that time period, the proportion of female deaths remained about 10 percent (Shankar and Varghese, 2006). Similarly, 92 percent of all motorcyclists injured in crashes are

Figure 4.1
Number of Motorcyclists Involved in Fatal Crashes by Age Group, 1998 and 2008

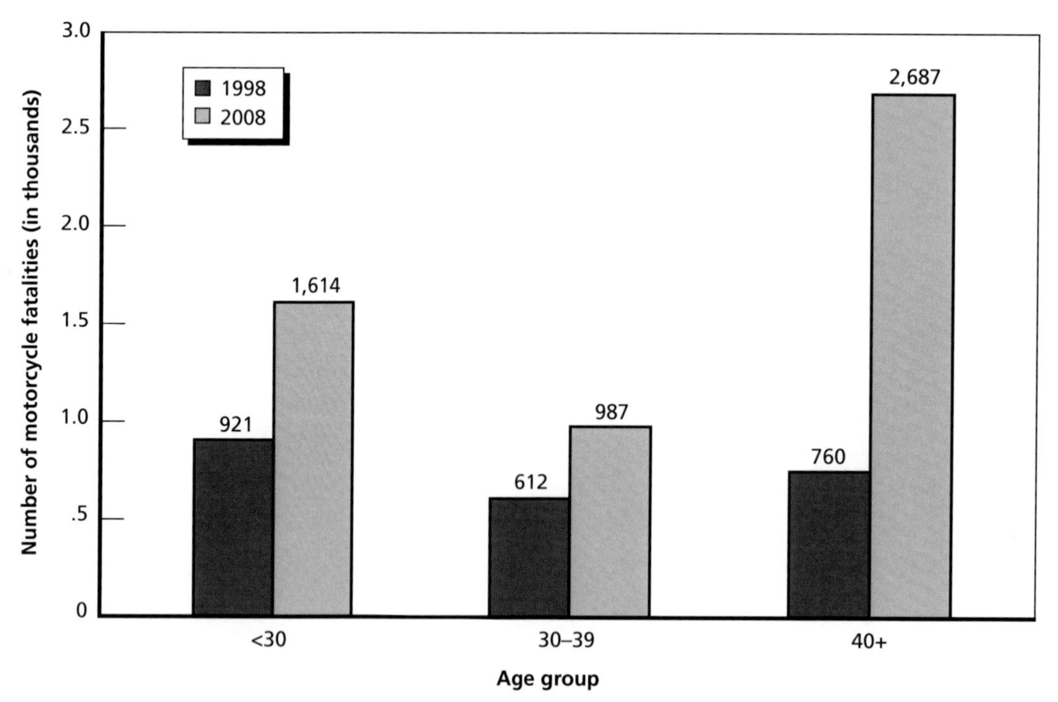

SOURCE: NHTSA, 2009a.
RAND *TR820-4.1*

motorcycle riders, as opposed to 8 percent who are motorcycle passengers (NHTSA, 2009b).[1] The injury trend is similar to the trend seen in motorcyclists killed in fatal crashes, but percentages of fatalities were not available (NHTSA, 2009b).

Since motorcycling is most popular in areas where riders can ride all year, fatalities are not evenly distributed geographically. Three states were responsible for about 30 percent of all motorcycle fatalities (4,955 total) in 2008: California (537 fatalities), Florida (523), and Texas (480) (NHTSA, 2009a). Florida has the highest per capita rate of motorcycle fatalities.

Socioeconomic Factors

Studies have found a moderate association between socioeconomic factors and motorcycle crashes. Hurt, Ouellet, and Thom (1981) reported that laborers, students, and unemployed individuals were more commonly involved in motorcycle crashes than professionals, sales workers, and craftsmen. There were also significantly more crashes for motorcycle riders that had not finished high school (Hurt, Ouellet, and Thom, 1981).

A 2006 study in Sweden found that among motorcycle riders 16 to 18 years old, those of lower socioeconomic status had higher odds of injury than their counterparts in the highest socioeconomic group (Zambon and Hasselberg, 2006). Specifically, they found that at age 16, motorcyclists in the lowest socioeconomic group had a risk of injury 2.5 times higher than those in the highest socioeconomic group. Some reasons offered to explain this difference include how much the motorcycle is used, where it is driven, the age and safety features of the motorcycle, motorcycle maintenance, access to driver training, availability and tendency to wear safety equipment, parental supervision, and familiarity with the vehicle. The results did not, however, account for possible differences in motorcycle use by socioeconomic status (Zambon and Hasselberg, 2006).

A 2001 study linked demographic data from FARS on fatal motorcycle crashes with lifestyle data from Claritas, a commercial geo-demographic database, to identify cost-effective crash prevention programs tailored to specific segments of the population (Shankar and Wardell, 2001). Claritas uses U.S. Census data to classify the 35,000+ zip codes into 62 clusters, each of which represents a unique set of demographic, socioeconomic and lifestyle characteristics. The authors found that younger riders in suburban areas and older riders in towns and rural areas have a higher incidence of fatal motorcycle crashes. Targeting urban and ethnic clusters and advertising through country music radio and television as well as motorcycle, fishing, and hunting magazines may be the best means to disseminate messages on motorcycle safety (Shankar and Wardell, 2001).

---

[1] In its 2007 traffic safety fact sheet on motorcycles, NHTSA redefined its motorcycle terminology as follows: "a **motorcycle rider** is the operator only; a **passenger** is any person seated on the motorcycle but not in control of the motorcycle; and any combined reference to the 'motorcycle rider' (operator) as well as the 'passenger' will be referred to as **motorcyclists**" (NHTSA, 2008a).

General Behavioral Factors

Wearing a Helmet

Motorcycle riders often sustain multiple injuries in a crash, of which lower-extremity injuries are the most common. However, head injuries are the most frequent injury in fatal crashes, contributing to over half of all deaths (Sosin, Sacks, and Holmgreen, 1990). Helmets and helmet use laws are effective in reducing head injuries and deaths from motorcycle crashes (Lin and Kraus, 2009). This subsection describes helmet use as a risk factor; helmet use laws are discussed in Chapter Seven.

NHTSA data on the effectiveness of helmets is conclusive. In 2008, NHTSA estimated that helmets saved the lives of 1,829 motorcyclists, and if all motorcyclists had worn helmets, an additional 823 lives could have been saved (NHTSA, 2009a). Helmets are estimated to be 37 percent effective in preventing fatal injuries to motorcycle riders and 41 percent for motorcycle passengers (NHTSA, 2009a). A 1996 NHTSA study using Crash Outcome Data Evaluation System data showed that motorcycle helmets are 67 percent effective in preventing traumatic brain injuries (NHTSA, 1996). A 2009 literature review of studies of helmet use in the United States found that "nonhelmeted riders are more likely to have head injuries, die, require longer hospitalization, and have higher medical costs compared to helmeted riders" (Lin and Kraus, 2009, p. 712). Note that these statistics refer to Department of Transportation (DOT)-compliant helmets; "novelty" helmets offer no protection against head injuries (NHTSA, 2007b).[2]

Helmet use by motorcyclists is not as high as seat belt use in cars. NHTSA National Occupant Protection Use Survey provides nationwide data on helmet use in the United States by observing motorcyclists at randomly selected roadway sites. In 2009, 67 percent of motorcyclists were wearing DOT-compliant helmets,[3] the highest proportion since 2000 (Figure 4.2). Helmet use rates for fatally injured motorcyclists have consistently tracked lower than overall helmet use, except from 2005 to 2007 (NHTSA, 2009a).

Helmet use by military personnel exceeds that of civilian motorcyclists. The 2008 Department of Defense Survey of Health Related Behaviors Among Active Duty Military Personnel (HRBS) found that 87 percent of military personnel who rode a motorcycle at least once in the past 12 months indicated that they "always" or "nearly always" wore a helmet when riding a motorcycle (Bray et al., 2009). Helmet use by military personnel has increased steadily since 1995 (Bray et al., 2009). However, these figures are self-reported as opposed to being derived from the National Occupant Protection Use Survey, which uses trained observers to collect information about helmet use. Therefore the results from these two surveys may not be directly comparable.

---

[2] All motorcycle helmets sold in the United States are required to meet NHTSA Federal Motor Vehicle Safety Standard (FMVSS) 218 (NHTSA, 2009a), which establishes the minimum performance requirements for helmets designed for use by motorcyclists. Helmets that meet FMVSS 218 are called DOT-compliant (or certified) and are identified by a sticker affixed to the helmet. Non-compliant helmets (sometimes called uncertified or "novelty" helmets) may not be sold as "motorcycle helmets." NHTSA tests of novelty helmets found significantly worse performance than DOT-compliant helmets in terms of their ability to absorb impact energy during a crash, prevent penetrating blows to the helmet, and remain strapped on the motorcyclist's head. The probability of brain injuries and skull fracture for motorcyclists wearing novelty helmets using computer simulations of head impact attenuation tests was 100 percent (NHTSA, 2007b). The report noted, "Motorcycle riders who wear novelty helmets and believe that 'something is better than nothing' have a false sense of security" (NHTSA, 2007b, p. 2).

[3] An additional 16 percent of motorcyclists were wearing novelty helmets, and the rest no helmets.

Figure 4.2
Motorcycle Helmet Use, 1995 to 2009

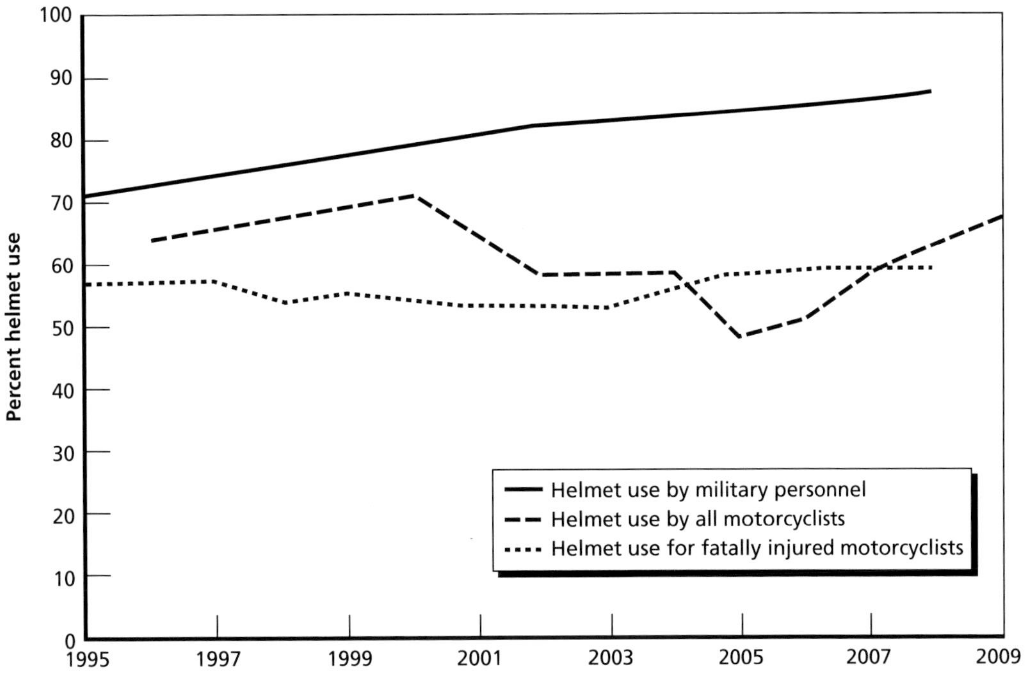

SOURCES: Bray et al., 2009; NHTSA, 2002, 2003, 2004, 2005a, 2006a, 2008b, and 2009a.
RAND *TR820-4.2*

Helmet use varies widely by state law and urban versus rural. In 2009, in states with helmet laws, 86 percent of motorcyclists wore helmets, whereas in states without helmet laws that figure was 55 percent (NCSA, 2009d). Helmet use in rural areas was 75 percent in 2009, while in urban areas it was 57 percent (NCSA, 2009d).

Use of Alcohol and Other Drugs

Alcohol use is a major factor in fatal motorcycle crashes to even a greater extent than for cars. In 2008, the most recent year for which statistics are available, almost 30 percent of motorcyclists killed had a BAC level of 0.08, compared with 23 percent of passenger car drivers. In addition, 43 percent of motorcycle riders who died in single-vehicle crashes had BAC levels of 0.08 or above (NHTSA, 2009a).

The proportion of motorcyclists killed who were intoxicated has declined since 1992 to 37 percent; see Figure 4.3. During this period, the proportion of motorcyclists killed with a BAC level below the legal limit of 0.08 has been between 7 and 9 percent, suggesting that even a few drinks can be fatal on a motorcycle.

For many years, motorcycle riders aged 15–29 years had the highest fatality rate in alcohol-related crashes (Lin and Kraus, 2009). However, for approximately the last 15 years, it has been older motorcycle riders who have had the highest alcohol-related fatality rate. In 2008, 41 percent of motorcyclists aged 45–49, 41 percent of motorcyclists aged 40–44, and 36 percent

Figure 4.3
Percentage of Motorcyclist Fatalities by BAC Level, 1992 to 2008

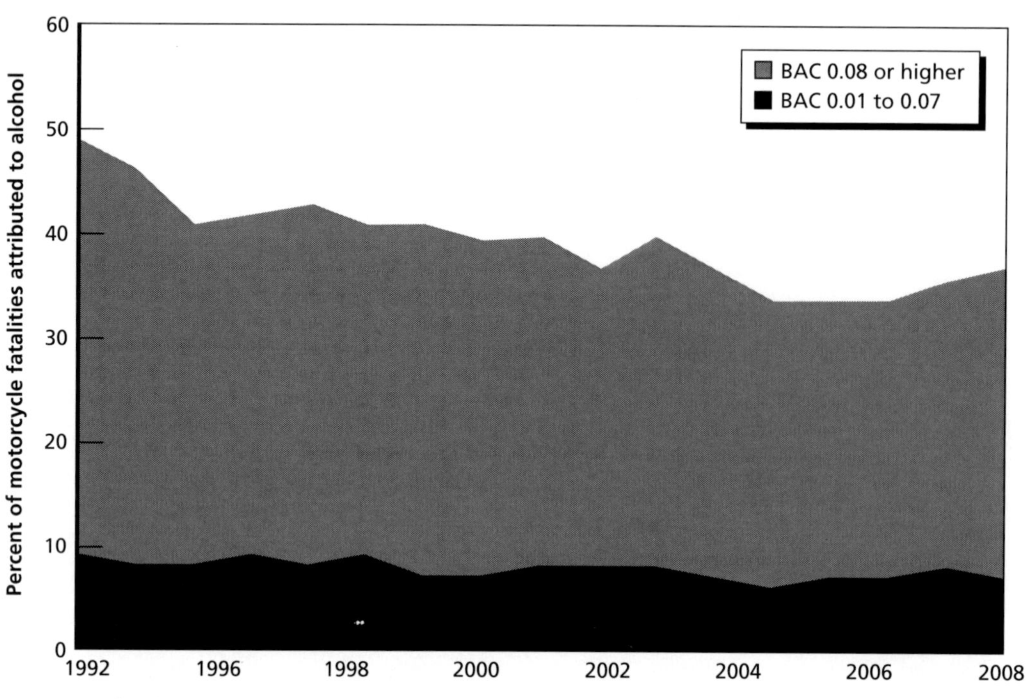

SOURCES: Shankar, 2003; NHTSA, 2002, 2004, 2005a, 2006b, 2007a, 2008b, 2009a.
RAND *TR820-4.3*

of motorcyclists aged 35–39 in fatal crashes had BAC levels of 0.08 or above (NHTSA, 2009a). This high rate of alcohol impairment by older motorcyclists along with increased licensing of and motorcycle ownership by older drivers appear to be contributing to the increase in fatal crashes observed in riders over the age of 40.

Motorcycle riders with BAC levels 0.08 or higher were more likely to be involved in fatal accidents at night (Williams et al., 1985; NHTSA, 2009a). In 2008, almost four times as many motorcycle riders with BAC levels over the legal limit were killed at night (48 percent) than during the day (13 percent) (NHTSA, 2009a). Motorcycle riders with BAC levels of 0.08 or higher involved in fatal crashes were also less likely to be wearing a helmet and to have more severe head injuries (NHTSA, 2009a; Lin and Kraus, 2009). Only 46 percent of motorcyclists killed with BAC levels of 0.08 or higher were wearing helmets, while 66 percent of motorcyclists who had no alcohol were wearing helmets (NHTSA, 2009a).

Two studies found that even small amounts of alcohol can impair motorcycle rider performance. The first study used a motorcycle simulator to test the performance of 14 experienced motorcycle riders with varying BAC levels. They found that motorcycle riders with BAC levels well below the legal limit had a tendency to leave the roadway and less ability to complete a timed course in the simulator test. At the legal limit,[4] even basic motorcycle handling skills were impaired (Colburn et al., 1993).

---

[4] The legal limit for BAC level at the time of the study was 0.10 in the state of Alabama where the study was conducted (Sun, Kahn, and Swan, 1998).

A later study used a closed test track to observe the effects of different BAC levels on experienced motorcycle riders' performance on a set of basic riding skills. They tested motorcycle riders at BAC levels of 0.00, 0.02, 0.05, and 0.08 on a test course that included basic riding skills such as offset weave (slalom), hazard avoidance, curve negotiation, and emergency stops.[5] Riders were impaired in motorcycle control and rider behavior at a BAC level of 0.08, but some effects were found at the lower BAC level of 0.05. The authors noted that because this study was conducted using experienced riders performing practiced skills on a closed test track, "larger impairments may be expected with less experienced riders on less familiar roads, with more complex and novel tasks at higher alcohol doses" (Creaser et al., 2007, p. 62).

Finally, motorcycle riders are more likely to have taken illegal drugs than car drivers. One study measured levels of marijuana, alcohol, cocaine, and phencyclidine (PCP) in automobile and motorcycle drivers treated at the Maryland Institute for Emergency Medical Services Systems Shock Trauma Center from June 1990 to March 1991. Marijuana use in motorcycle riders involved in crashes was 12 times higher than in automobile drivers (32 percent vs. 2.7 percent). In addition, more motorcycle riders than automobile drivers had used alcohol (47.1 vs. 35.2 percent), cocaine (8 vs. 5 percent), and PCP (3.1 vs. 1.5 percent) (Soderstrom et al., 1995). A previous study of patients at the Maryland Institute for Emergency Medical Services Systems Shock Trauma Center found that almost as many motorcycle riders who had been injured had used both alcohol and marijuana than those that used alcohol alone (24.3 vs. 25.7 percent, respectively) (Soderstrom et al., 1988). Another study found that 40 percent of fatally injured motorcycle riders aged 15 to 34 had used combinations of alcohol, marijuana, and/or cocaine as compared to 34 percent who had only used one drug (Williams et al., 1985).[6]

Speeding

Speeding is a major factor in motorcycle crashes (NHTSA, 2003). In particular, inappropriate speed for the road or traffic conditions and excessive speed increase the risk of causing a motorcycle accident (Lardelli-Claret et al., 2005). In addition, motorcyclists traveling at higher speeds at the time of impact tend to have more serious injuries (Lin and Kraus, 2009).

In 2008, NHTSA reported that 35 percent of all motorcycle riders involved in fatal crashes were speeding, compared with 23 percent for passenger car drivers (NHTSA, 2009a).[7] Almost two thirds of the fatalities in single-vehicle motorcycle crashes were associated with speeding (Shankar, 2001).

Few studies look at the effect of speed limits specifically on motorcycle fatalities (Lin and Kraus, 2009). One longitudinal analysis of state-specific data from 1990 to 2005 to determine how various alcohol and traffic policies impact motorcycle safety found that a 10 mile per hour (mph) reduction in the speed limit had no effect on the fatal injury rate and actually increased the nonfatal injury rate by about 11 percent (French, Gumus, and Homer, 2009). A study in South Wales of mobile speed cameras found a 63 percent decrease in motorcycle crashes caus-

---

[5] The test track was developed based on standard exercises in MSF's training program (Creaser et al., 2007).

[6] This study defined alcohol as a drug.

[7] NHTSA considers a crash to be speeding-related if the driver was charged with a speeding-related offense or if a police officer indicated that racing, driving too fast for conditions, or exceeding the posted speed limit was a contributing factor in the crash (NHTSA, 2009a).

ing injuries at camera sites, including crashes occurring in daytime and nighttime, and on roads with speed limits of 30 and 60–70 mph (Christie et al., 2003).

Licensing, Ownership, and Experience

All states and the District of Columbia require motorcycle riders who use public roads to have a valid motorcycle license or endorsement (NHTSA, 2006a). Riding a motorcycle without a valid license is associated with higher risks of crashing and serious motorcycle injury (Kraus et al., 1991; Lardelli-Claret, 2005; Lin and Kraus, 2009). In 2008, 25 percent of motorcycle riders involved in fatal crashes were not licensed or were improperly licensed at the time of the crash (NHTSA, 2009a). In comparison, only 12 percent of drivers of passenger vehicles in fatal crashes did not have valid licenses (NHTSA, 2009a). In addition, 18 percent of motorcycle riders involved in fatal traffic crashes had previous license suspensions or revocations as compared to 13 percent of passenger vehicle drivers (NHTSA, 2009a). Young drivers under the age of 20 have the lowest licensure rates, while drivers age 40 and older had the highest rate of licensure (Kraus et al., 1991).

A 1991 study found that, at all ages, owners of motorcycles involved in crashes were less likely to have valid licenses than motorcycle owners not in crashes. In addition, motorcycle drivers who crashed and who did not own the motorcycle they were riding were more likely to be unlicensed than those owning the motorcycle. Almost all of the young motorcyclists who crashed on motorcycles that were owned by other young and unlicensed motorcyclists had invalid licenses (Kraus et al., 1991). A later report concludes that, "lack of a license, ownership, and youth are correlated, and all of these factors are associated with higher risks of motorcycle crashes and injuries" (Lin and Kraus, 2009, p. 717).

While several studies showed an association between less driving experience and a higher risk of motorcycle crashes and injuries (Mullin et al., 2000; Lin and Kraus, 2009), when controlling for other variables, youth is a bigger predictor than inexperience of motorcycle crashes and injuries (Rutter and Quine, 1996; Mullin et al., 2000). In a national prospective survey of 4,101 riders in the UK, Rutter and Quine (1996) found that youth played a greater role than inexperience through a pattern of risk-taking behaviors, in particular a willingness to break the law and violate the rules of safe riding. Mullin et al. (2000) found that familiarity with the specific motorcycle being ridden was the only type of experience that was associated with a strong protective effect.

Conspicuity of the Motorcycle and Rider

Conspicuity, the ability of motorcyclists to be seen easily by drivers of other vehicles, is an important factor in preventing motorcycle crashes. However, two-thirds of car drivers claimed not to have seen the motorcycle or to have seen it too late to have avoided a collision, and motorcyclists claim "that when they ride they become invisible" (Hurt, Ouellet, and Thom, 1981; Blanchard and Tabloski, 2006, p. 331). Several ways to increase the conspicuity of motorcycles and riders have been studied, including DRLs,[8] colored lights and modulators; color of the motorcycle; reflective, fluorescent or brightly colored clothing; and helmet color (Hurt, Ouellet, and Thom, 1981; NHTSA and MSF, 2000; Wells et al., 2004; Blanchard and Tabloski, 2006; Lin and Kraus, 2009).

---

[8] Daytime running lights (DRLs) are lights on the front of vehicles that are intended to improve the conspicuity of the vehicles during daylight by providing enough light to contrast the vehicle from its background.

Federal law does not require motorcycles or other passenger vehicles (i.e., passenger cars and light trucks and vans) to have DRLs, although all manufacturers voluntarily equip motorcycles with such lights (Wang, 2008). Twenty-five states require motorcyclists to ride with their headlights on at all hours.[9] In addition, several countries have laws requiring the use of DRLs on motorcycles, including Canada, Australia, Denmark, Germany, France, and Spain (Rumar, 2003). These laws are based on several studies that found the use of DRLs by motorcyclists decreased the risk of motorcycle crashes by increasing conspicuity (Hurt, Ouellet, and Thom, 1981; Henderson et al., 1983; Zador, 1985; Wells et al., 2004; NHTSA, 2006a). Other studies have found that DRLs do not reduce crashes between motorcycles and other vehicles (Muller, 1982; Paine et al., 2005). Some have also argued that the increasing use of DRLs in passenger vehicles negates the positive effects of motorcycles operating with their lights on (NHTSA and MSF, 2000; NHTSA, 2006a; Lin and Kraus, 2009). Alternative DRL technologies proposed to increase motorcycle conspicuity include modulating headlights that cause the light to alternate between a higher and a lower intensity during the day,[10] turn signal DRLs, yellow or amber DRLs, and additional lights located near the existing DRL to create a unique DRL configuration for motorcycles (NHTSA and MSF, 2000; Rumar, 2003; Paine et al., 2005).

The color of a motorcycle, helmet, and clothing worn by the rider can also play a role in conspicuity. Motorcycles with brightly colored fairings (a covering placed over the frame that reduces air drag and protects riders from wind and weather) appeared to be effective in increasing conspicuity and preventing crashes (Hurt, Ouellet, and Thom, 1981). Helmet color has also been discussed with respect to motorcycle conspicuity and risk of crashing. Wells et al. (2004) report that wearing light colored helmets (such as white) is associated with increased conspicuity and reduced risk of crashing. However, Hurt, Ouellet, and Thom (1981) reported no association between helmet color and increased conspicuity. In addition, studies have also shown that motorcyclists wearing reflective, fluorescent or brightly colored clothing are more visible and were at a lower risk of being injured in a crash (Hurt, Ouellet, and Thom, 1981; Kwan and Mapstone, 2004; Wells et al., 2004; NHTSA and MSF, 2000).

Increased conspicuity can help other drivers see motorcyclists, which is desirable since many crashes are the result of other drivers not seeing motorcyclists. In 2008, 47 percent of all fatal motorcycle accidents involved crashes between motorcycles and another type of moving motor vehicle (NHTSA, 2009a). According to the National Agenda for Motorcycle Safety, "when motorcycles and other vehicles collide, it is usually the other (non-motorcycle) driver who violates the motorcyclist's right-of-way" (NHTSA and MSF, 2000). Motorcycles are obviously smaller than cars and there are far fewer on the roads, and drivers of other motor vehicles may not be used to seeing them and dealing with them in traffic (NHTSA and MSF, 2000). In addition to making riders and their motorcycles more visible, it is also important to educate other motorists to increase their awareness of motorcycles (Baer and Skerner, 2009; NHTSA, 2006a).

---

[9] The 25 states that require daytime use of headlights are Alaska, Arkansas, California, Connecticut, Florida, Georgia, Illinois, Indiana, Iowa, Kansas, Maine, Minnesota, Montana, New York, North Carolina, Oklahoma, Oregon, Pennsylvania, South Carolina, Tennessee, Texas, Washington, West Virginia, Wisconsin, and Wyoming (American Motorcyclist Association, 2010).

[10] Modulating headlights are permitted in all 50 states per federal regulation 49 CFR 571.108 S7.9.4 (NHTSA and MSF, 2000).

Engine Size

Of particular interest to public health and safety experts are sport and supersport bikes, lightweight motorcycles with large powerful engines capable of quick acceleration and high speeds, which have become quite popular over the last ten years or so, especially with riders younger than 30 years of age (Insurance Institute for Highway Safety [IIHS], 2007; Morris, 2009). Sales of sport bikes and supersport bikes increased from 16 to 19 percent of all motorcycle sales between 2005 and 2007 (Morris, 2009). Sales of motorcycles with bigger engine sizes (750 cubic centimeters [cc] or more) account for about 75 percent of all motorcycle sales. Sales of motorcycles of all engine sizes have more than doubled from 1998 to 2003.[11] Table 4.1 describes the different types of motorcycles.

Since the late 1990s, the number of fatalities on motorcycles with larger engines has grown significantly (see Figure 4.4) However, the percentage of fatalities by engine size has remained roughly constant, suggesting that the increasing engine size of sport and supersport bikes is not the sole reason for the growth in motorcycle fatalities.

Table 4.1
Types of Motorcycles

Type	Description	Engine Size (cc)	Weight (pounds)	Average Rider Age	
				IIHS	MIC
Cruiser	Largest class of bikes; emphasis on appearance, style, and sound; less emphasis on performance; long profile with low saddle height; often customized	650–1,800	700–1,300	45	44
Standard	Have basic designs and upright riding positions; have low power-to-weight ratios; considered to be user-friendly motorcycles	125–1,800	200–1,200	39	44
Supersport	Consumer versions of racing motorcycles; reduced weight and increased power allow for extreme acceleration and speeds of nearly 190 miles per hour	650–1,100	290–350	33	31[b]
Sport	Styled and built in the manner of road-racing motorcycles with a forward leaning riding position; capable of high speeds but do not have the acceleration, stability, and handling of supersports; emphasis on handling, acceleration, speed, braking, and cornering	650–1,100	290–350	39	31[b]
Unclad Sport	Similar to sport motorcycles and supersports in design and performance but without plastic body fairings	650–1,100	290–350	38	31[b]
Touring	Have big engines and fuel tanks plus room to haul luggage; often outfitted with antilock brakes, audio systems, and cruise control	1,600–1,800	800–950	mid-40s	52
Scooter	Have small wheels, automatic transmissions, and small engines; larger scooters are becoming more popular	50–650	220–500[a]	48	40

SOURCES: IIHS, 2007; Morris, 2009; Charlie Fernandez, email to Liisa Ecola on May 6, 2010, regarding "MIC [Motorcycle Industry Council] Review of: Understanding and Reducing Off-Duty Vehicle Crashes Among Military Personnel."

[a] Chappell, 2008.

[b] Average for all sport bike types.

---

[11] Charlie Fernandez, email to Liisa Ecola on May 6, 2010, regarding "MIC [Motorcycle Industry Council] Review of: Understanding and Reducing Off-Duty Vehicle Crashes Among Military Personnel."

Figure 4.4
Motorcyclist Fatalities by Engine Size (cc), 1998 to 2008

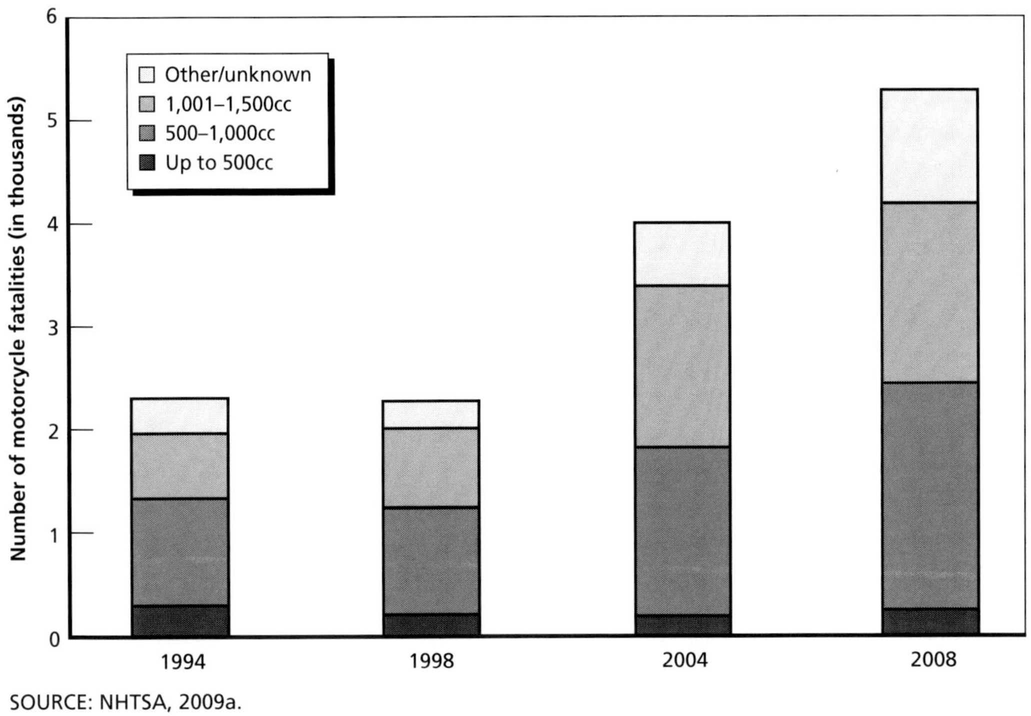

SOURCE: NHTSA, 2009a.
RAND *TR820-4.4*

In 2007, an IIHS report found that "motorcyclists who ride supersports have driver death rates per 10,000 registered motorcycles nearly 4 times higher than motorcyclists who ride all other types of bikes" (IIHS, 2007). Of all motorcycle riders killed in 2005, those riding supersports were the youngest, with an average age of 27, and touring motorcyclists were the oldest with an average age of 51. The average age for fatally injured motorcycle riders of sport bikes was 34 (IIHS, 2007).

Most fatal crashes on sport and supersport bikes were caused by speeding and driver error. Speed was a factor in 57 percent of supersport riders' fatal crashes in 2005 and 46 percent of the fatal crashes of sport riders. In comparison, speed was a factor in 27 percent of fatal crashes among riders of cruisers and standards and 22 percent on touring motorcycles. Helmets were not enough to overcome the problem of speeding; 71 percent of supersport riders who died in crashes in 2005 were wearing helmets, compared with 52 percent of touring motorcyclists (IIHS, 2007).

Service Safety Center briefs to the PMV TF have indicated that drinking-related motorcycle mishaps appear to be lower in the military sport-bike population than with their civilian counterparts, but this conclusion appears to be anecdotal at this point.

CHAPTER FIVE

Predicting Risky Driving

Risky driving[1] commonly refers to driving while intoxicated (DWI),[2] failure to use seat belts, frequent/excessive speeding, aggressive driving (e.g. following closely, cutting people off), and failure to observe traffic signals, signs or rules. Understanding which drivers are most likely to take risks behind the wheel is key to developing effective safety programs.

Looking at risky driving as a cluster of behaviors, instead of analyzing each one separately, is logical both conceptually and empirically. While some studies, particularly of DWI and seat belt use, look at only one behavior, many studies include multiple forms of risky driving in a single analysis, creating summed scores indicating the number or frequency of engaging in any or all of them. The factors associated with each aspect of risky driving, as well as the behaviors themselves, overlap considerably. For example, gender, age (Abdel-Aty and Abdelwahab, 2000; Caetano and Clark, 2000), and binge or heavy drinking (Bray and Hourani, 2007) are commonly reported correlates of DWI. Reviewing intersection crashes in both FARS and the General Estimates System during a five-year period, Retting, Ulmer, and Williams (1999) found that red light runners involved in crashes were more likely than other drivers to be younger than age 30, to be male, to have convictions for DWI, and to have consumed alcohol prior to the crash. They were also more likely to have prior moving violations. Similarly, younger drivers are less likely to wear seat belts, as are drinking drivers. Unrestrained drivers killed in 2003 were twice as likely to have a positive BAC level as restrained drivers (Subramanian, 2005). Nichols, Chaudhary, and Tison (2009) note that most people who die in alcohol-related crashes are not buckled up. Almost two-thirds of young drivers in alcohol involved crashes were not wearing belts, and 77 percent of young drivers in fatal alcohol involved crashes were not belted (NHTSA 2008a).

Given the strong associations observed among risky driving behaviors, we review the predictors of these behaviors as a group, noting the particular behavior that was examined in cases where a specific form of risky driving was studied, and referring to "risky driving" as a general outcome in cases where multiple risky driving behaviors were combined in a survey or analysis.

---

[1] The public health literature tends to discuss "risk behavior" while the traffic literature uses the term "risky driving." For consistency, we refer to both risky behavior and risky driving.

[2] Driving while intoxicated (DWI) and driving under the influence (DUI) mean the same thing legally; different phrases are used in different states. For consistency, we use DWI in this report even if the authors of a study used DUI.

Age and Gender

As noted throughout the preceding chapters, several forms of risky driving behavior are more common among young people, particularly young men: DWI (Abdel-Aty and Abdelwahab 2000; Caetano and Clark, 2000), failure to use seat belts (Fell et al., 2008; Vivoda et al., 2007), speeding over 10 mph above speed limits, and speeding for the thrill of it (Oltedal and Rundmo, 2006; Blows et al., 2005; Ryb et al., 2006; Sabel, Bensley, and van Eenwyk, 2004), following too closely to the vehicle ahead (Fergusson, Swain-Campbell, and Horwood, 2003), violations of signals and signs such as running stop lights (Retting, Ulmer, and Williams, 1999) and a variety of other offenses (Williams, 1998). Thus, the preponderance of young men in the military population renders personnel predisposed to risk.

Heavy, Binge, and Problem Drinking

A key predictor of risky driving in general, and DWI in particular, is excessive drinking. While definitions of binge drinking, heavy drinking, and problem drinking vary by researcher, they generally refer to the following:

- Binge drinking: for men, consuming five or more drinks in one sitting, or four drinks for women
- Heavy drinking: binge drinking at least once a week, or having at least 14 drinks a week for men or seven for women
- Problem drinking: drinking alcohol in a manner that creates other problems, such as with job performance.

Just as military personnel are more likely to be young and male, they are also more likely to be heavy or binge drinkers. Indeed, the percentage of active-duty military personnel (ADMP) who are heavy or binge drinkers is substantially higher than the percentage observed in civilian populations of similar age and gender (Ames and Cunradi, 2004/2005). The following are some key statistics:

- In 2008, 28.4 percent of military males aged 18–25 reported binge drinking during the past month, compared with 17.9 percent of civilian males the same age (Bray et al., 2009).
- Among females that age, the respective military and civilian rates were 10.7 and 7.3 percent (Bray et al., 2009).
- Forty-seven of all ADMP reported binge drinking in the past month (Bray et al., 2009).
- Heavy drinkers (19.8 percent of ADMP) were responsible for 71.5 percent of the binge-drinking episodes (Stahre et al., 2009).
- 67.1 percent of binge-drinking episodes were reported by personnel aged 17–25 years (46.7 percent of ADMP), and 25.1 percent of these episodes were reported by underage youth (aged 17 to 20 years), who make up 14.1 percent of ADMP (Stahre et al., 2009).

Heavy and binge drinking have been increasing in the military over the past decade. For example, in 1998 34.9 percent of all Department of Defense (DoD) personnel were considered binge drinkers; by 2008, this figure was 47.1 percent (Bray et al., 2009). These trends were not attributable to demographic shifts in the composition of ADMP over this period (Bray and Hourani, 2007). On the other hand, illicit drug use has been declining.

Predictors of drinking in ADMP include being male, younger, less educated, divorced or single, and white (Ames and Cunradi, 2004/2005; Bray and Hourani, 2007; Jacobson et al., 2008). Across different branches of the military, rates of heavy drinking are highest among Marines and Army males, exceeding those among civilians after adjusting for demographic differences. Rates among Navy males are similar to adjusted civilian rates as are all rates for females in services other than the Marine Corps. Heavy drinking among Air Force men is lower than among demographically similar civilians (Ames and Cunradi, 2004/2005; Bray and Hourani, 2007). Problems from drinking appear to be more common among National Guard and Reserve troops than among active duty personnel (Jacobson et al., 2008). Personality predictors of drinking documented among reserve and guard members about to deploy include "negative emotionality" (a tendency to be emotionally reactive and anxious, perceiving everyday experiences as threatening) and "disconstraint" (a tendency to act impulsively or without regard to norms) (Ferrier-Auerbach et al., 2009).

Heavy drinking clearly predicts drinking and driving among military personnel. Among ADMP participating in the 2005 HRBS, the rate of self-reported driving "a car or another vehicle after having too much to drink" was 33.2 percent in heavy drinkers and only 7 percent among light or infrequent drinkers. Thirty-eight percent of heavy drinkers reported that they rode as a passenger with a person who had too much to drink (Bray and Hourani, 2007). The 2008 update of this survey did not report on the link between heavy drinking and drinking and driving, and neither report discussed whether binge drinking similarly predicted drinking and driving.

A separate analysis of the same data showed that binge drinkers in the military are commonly underage (less than 21 years) and that the joint problems of heavy or binge drinking and underage drinking are strongly associated with alcohol-impaired driving in the military (Stahre et al., 2009). In a study using the military Health Risk Assessment surveys from 1990–1998 and looking exclusively at male respondents, it was found that high-risk drinkers were both less likely to wear seat belts and more likely to speed (>15 mph over limit) (Williams, Bell, and Amoroso, 2002). Problem drinking may also contribute to DWI in the military. In a sample made up of ADMP, guard and reserve members, those who had previously been hospitalized with a diagnosis of substance abuse were more likely to die in a subsequent motor vehicle crash (Hooper et al., 2006). However, it is not reported whether alcohol or drug use contributed directly to the crashes that were studied.

We obtained data from two services, the Navy and Marines, about the prevalence of drunk driving as a cause of crash deaths for both passenger vehicles and motorcycles. Table 5.1 shows these figures. Given that about one-third of U.S. vehicle fatalities are the result of drunk driving, the Navy is slightly higher than the U.S. total, while the Marines are below.

Interestingly, the proportion of deaths attributable to alcohol is lower in the services for motorcycles than for cars and trucks. This is the reverse of the patterns in the U.S. population, where the proportion of motorcyclists in fatal crashes driving with a BAC level of 0.08 is higher than that of car and truck drivers—over this time period, the percentage of car/truck drivers with a BAC level over 0.08 is generally around 22–23 percent, while for motorcyclists is it about 27–28 percent.[3]

These figures do not prove that the high rates of heavy and binge drinking in the military produce a high number of drunk driving crashes. However, they suggest two things. First, drunk driving is a significant factor in military crash deaths. The Navy and Marine data do not include other causes of crashes, so it is not clear if alcohol is the single largest factor. Second, the impacts of drunk driving are much higher for cars and trucks than for motorcycles. This seems to corroborate the finding in Bowes and Hiatt (2008) that risks differ between car drivers and motorcyclists.

The data we have do not allow us to examine more closely drunk driving by age in the military. For the U.S. population, the percentage of crashes attributable to drunk driving is highest among those people their 20s, but lower for teenagers. It is not clear if drunk driving fatalities in the military follow a similar pattern.

It is possible that one reason rates of heavy and binge drinking are so high is that military culture promotes drinking. The 2005 HRBS found that 20 percent of infrequent or light drinkers believe that "drinking is part of being in the military," and nearly double that percentage, 39 percent of heavy drinkers, agreed. Many military personnel certainly believe that military culture at least does not disapprove of drinking, with similar percentages of light and heavy drinkers, 18 and 37 percent, reporting that "leadership is tolerant of off-duty drinking" (Bray and Hourani, 2007). Changing the culture surrounding drinking in the military might reduce traffic deaths by reducing heavy drinking.

Stress may also contribute to drinking among military personnel; anecdotally, drinking is tolerated in the military as a way of allowing personnel to "blow off steam." Substance abuse is highly correlated with PTSD, a common response to combat deployment (McFall, Mackay, and Donovan, 1992). Combat deployment is associated with increased rates of heavy, binge, and problem drinking, particularly among reserve and guard personnel (Jacobson et al., 2008). But high levels of everyday stress may also contribute to heavy drinking among military personnel. Subjective reports by military men of "a great deal" or "a fairly large" amount of work-related stress in response to the 1995 HRBS were associated with their reports of heavy drinking (odds ratio of 1.4 compared with those reporting no work-related stress). Women's reports of work-related

Table 5.1
Incidence of Drunk Driving as a Cause of Motor Vehicle Crash Fatalities, 2002–2009

	Average Percentage of Crash Fatalities Attributed to Alcohol Use by Drivers	
	Cars/Trucks	Motorcycles
Navy	39	23
Marines	31	12

SOURCE: Naval Safety Center, 2010.

[3] We do not show U.S. figures in the table since the reports from which we compiled these figures (NHTSA motorcycle *Traffic Safety Facts* for various years) compare the percentage of *drivers* with BAC over 0.08, not the number of *people* killed in crashes caused by such drivers. Therefore, these figures are not directly comparable to the military figures.

stress did not predict heavy drinking. The authors argued that women cope with stress in different ways than men, and that this accounts for the findings. However, heavy drinking was quite rare among military women (5 percent of women vs. 19 percent of men reported engaging in this behavior). Thus, these reported results may also reflect the limitations of statistical tests for predicting rare behaviors (Bray, Fairbank, and Marsden, 1999).

Personality

While demographic and cultural factors help explain why certain groups drink more heavily, personality factors are also linked with risky driving. This section looks at three such factors:

- Sensation-seeking is a trait "defined by the seeking of varied, novel, complex and intense sensations and experiences and the willingness to take physical, social, legal and financial risks for the sake of such experiences" (Zuckerman, 1994, p. 27).
- Impulsivity is a tendency to act without thinking or failure to inhibit risky behavior (Zuckerman, 2007).
- Deviance is frequent antisocial behavior, such as lying, cheating on a test, minor theft, etc. (Jessor and Jessor, 1977).

A number of studies link sensation-seeking and risky driving, particularly a subcomponent of sensation-seeking labeled "thrill seeking" (Jonah, 1997; Zuckerman, 2007). Most of these studies are based on surveys that link scores on measures of sensation-seeking with self-reported risky driving behavior and self-reported traffic violations, including DWI (Jonah, 1997; Zuckerman, 2007). However, some studies have directly observed driving speed and other behaviors (e.g., Heino, van der Molen, and Wilde, 1996; Burns and Wilde, 1995). Diverse samples have been studied, including groups in the United States, Canada, Great Britain, the Netherlands, Sweden, Finland, and Norway. Many studies are based on convenience samples of college students, but others include large samples of adult drivers, and among studies examining drunk driving, a large percentage include samples of drivers arrested and/or convicted of DWI. This wide variety of evidence in diverse groups provides strong support for the role of sensation-seeking in risky driving. Moreover, in a review of this literature, Jonah (1997) found correlations in the 0.30 to 0.40 range, indicating that sensation-seeking personality can account for a substantial portion, up to 16 percent, of the variation in drivers' risky behavior.

Zuckerman (2007) argues that sensation-seeking predicts risky driving primarily because sensation-seekers enjoy the sensation of speeding, and not being able to speed results in other risky driving tactics (e.g., following closely and lane changing). Sensation-seekers desire not risk, but speed. He notes in reference to a study by Heino, van der Molen, and Wilde (1996) that sensation-seekers can be divided into high and low sensation-seekers based on how they think about risk. Both accept the same level of risk, but low sensation-seekers see certain actions as more risky. For example, both low and high sensation-seekers might not speed unless they think the chance of being in a crash is less than 5 percent, but a low sensation-seeker might think a speed of 10 mph over the limit creates a 5 percent risk, while a high sensation-seeker might think 20 mph creates a 5 percent risk. It may also be that high sensation-seekers simply value the experience of speed more highly. That is, they recognize the risks of speeding, and would prefer to avoid these risks as much as would low sensation-seekers, but their strong desire for speed outweighs their desire to limit risks. This view of sensation-seeking as the pursuit of sensory experience, rather than the pursuit of danger, applies to other aspects of risky driving as well as speed, but Zuckerman considers speed most central in understanding the association between driving and personality. It may be for this reason that evidence linking sensation-seeking to seat belt use is mixed, with two studies finding an association (Thuen, 1994; Wilson, 1990) and one not (Jonah, Thiessen, and Au-Yeung, 2001). Seat belt use may have little effect on the sensations one experiences while driving, and it has a large effect on driving risk.

Impulsivity is particularly characteristic of youth (recent research links the impulsivity of youth to a stage of brain development; see Steinberg, 2009, for a review). A review of 11 studies (Araujo, Malloy-Diniz, and Rocha, 2009) linked impulsivity to driver violations, speeding for thrill, and traffic accidents.

Sensation-seeking and impulsivity have been repeatedly linked to risky driving in the general population, and to drinking and driving in general and military populations, although their relationship to drinking per se only partially explains these relationships. Sensation-seeking also appears to directly affect DWI and other risky driving behaviors (Cherpitel, 1999; Yu and Williford, 1993; Zuckerman, 2007). Sensation-seeking is a well-established correlate of drinking in the general population (Zuckerman, 2007), although it has not been specifically tested as a predictor of drinking among military personnel.

The 2008 HRBS confirms that these relationships hold true for service members (Bray et al., 2009). Analyses of these data find that impulsivity and sensation-seeking among military personnel are related to their seat belt use, drinking (any consumption of alcohol in the past 30 days whether driving or not), heavy drinking, driving after drinking too much, and other substance use. Differences on the drinking and driving item were pronounced, with 13.6 percent of those classified as high in impulsivity versus 3 percent of those scoring low on the measure reporting driving after drinking too much. On the measure of sensation-seeking, 10.9 percent of high sensation-seekers drove after drinking, versus 3.5 percent of low sensation-seekers.

The same report includes percentages of DoD personnel classified as low, moderate, and high on impulsivity and sensation-seeking, using a scale and scoring system identical to the National Alcohol Survey, a general population study of U.S. adults conducted in 1995 (Cherpitel, 1999). Table 5.2 compares how the military and civilian populations rank on these surveys.

For both personality factors, the military had a far higher proportion of personnel who scored high. Higher rates of these characteristics among young men can explain at least a portion of these differences, but the data indicate clearly that one reason for high rates of risky

Table 5.2
Rates of High, Medium, and Low Sensation-Seeking and Impulsivity in Military and Civilian Populations

	Sensation-Seeking (%)		Impulsivity (%)	
	Military	Civilian	Military	Civilian
High	78	27	46	7
Medium	19	38	46	33
Low	3	35	8	60

SOURCES: Bray et al., 2009; Cherpitel, 1999.

driving in the military is the presence of more risk-prone personality types among military personnel.

A third personality factor that has been linked to risky driving, including DWI in a military sample (Lucker et al., 1991) is the so-called "deviance-prone personality" (Bingham and Shope, 2004a, 2004b). In literature addressing risky behaviors such as substance use and violence among adolescents, Problem Behavior Theory (Jessor and Jessor, 1977) argues that a constellation of risky, anti-social behaviors stems from certain backgrounds and environments. In their review of the literature exploring individual differences in crash risk, Elander, West, and French (1993) concluded that risky driving is an expression of antisocial personality, spanning the range from mild social deviance to disorder. Their conclusion was based on studies linking speeding, crash rates, and other types of risky driving behavior to characteristics like emotionality, hostility, impulsivity, aggression, and seeking of authority,[4] as well as associations between crash rates and criminal offenses and deaths by violent act. The patterns of association between risky driving and other forms of deviance in adolescence and young adulthood (such as heavy drinking) and the similarity in predictors of these behaviors, provides support for considering risky driving a manifestation of deviance-prone personality (see Jessor and Jessor, 1977). However, the wide array of attributes that Elander, West, and French (1993) argue are reflective of this disposition suggest that other individual differences may contribute to risky driving and/or crash risk, in addition to social deviance.

Although they have received little attention, a few other traits have also been linked to risky driving in the general population. They are mentioned here because they may suggest levers for altering the tendency to drive unsafely, as well as individuals to target with interventions. First, there is some evidence that more aggressive individuals are more risky drivers (Arnett, Offer, and Fine, 1997; Matthews et al., 1999; Patil et al., 2006; Zuckerman and Kuhlman, 2000). Traits that have been linked to *low* rates of risky driving are forgiveness, consideration of future consequences (Moore and Dahlen, 2008), and altruism (Machin and Sankey, 2008). It is possible that appealing to these feelings in traffic safety campaigns might, therefore, be an effective means of reducing risky behavior.

Beliefs

Risk perception is the most commonly studied belief in the literature on risky driving. Studies consistently demonstrate that those who drive less safely also rate the risks of such driving behavior as low (e.g., Ryb et al., 2006; Ulleberg and Rundmo, 2003).

For example, Castella and Perez (2004) asked participants to indicate which of 28 risky driving behaviors they engaged in and to rate the level of risk posed by each. The correlation between scores on these two scales was 0.36 among men and 0.38 among women, indicating a moderate relationship between perceptions of risk and risk-taking. Similarly, Gullone et al. (2000) found that scores on a measure of "reckless behavior" that consisted almost entirely of risky driving behaviors (e.g., drinking and driving, speeding, joy riding, and driving without a license) showed these behaviors as less risky. Consistent with Castella and Perez's results, correlations were moderate in size and ranged from –0.20 to –0.38.

---

[4] *Seeking of authority* refers to trying to gain authority over others.

Although some have used these patterns to argue that risk perceptions cause risky driving, these associations may be driven by self-justification—those who drive dangerously may describe their behavior as less risky in order to feel better about it. A unique two-year study found that risk perception did not add to the investigators' ability to explain crash risk beyond what they could learn from drivers' behavior. Ivers et al. (2009) tracked more than 20,000 young Australian drivers over a two-year period. Driving records were linked with survey responses at the study's outset. Those in the top third of risky driving behaviors were 50 percent more likely to be involved in a later crash, but perceptions of a variety of driving practices as risky were not associated with the risk of later crash once baseline driving practices were also accounted for. These data suggest that crashes have little basis in drivers' risk perceptions. However, the lack of other longitudinal studies tracking relationships between risk perceptions and subsequent driving behavior and outcomes makes it difficult to determine their importance definitively (Irwin and Halpern-Felsher, 2002).

Nonetheless, the presumed importance of risk perception to behavior is well-grounded in psychological theory. All of the well-validated models explaining why people engage in risky or risk-reducing behavior recognize the importance of perceived risk, or more generally, perceived consequences. Yet no theory argues that risk perception is the only belief that drives risky behavior, and few identify it as the most important. Given this, it is surprising how little work has examined other well-established cognitive predictors of risky behavior. These include perceptions of social norms (e.g., do other people engage in the risky behavior, take steps to reduce risk, and/or approve or disapprove of those who do so?), and perceptions of self-efficacy (perceptions of one's own ability to engage in protective action or refrain from engaging in risk). Such factors not only predict risky behavior, but have played a central role in effective interventions to change these behaviors.

We identified only one study of risky driving that explored the importance of this broader set of predictors. It examined the Theory of Planned Behavior (Fishbein and Ajzen, 1975), a well-validated theory of deliberate action that has been applied to predict and explain multiple forms of risky behavior. The Theory of Planned Behavior argues that behavior is a direct function of intentions, and that intentions are a result of three factors: perceived norms, perceived control over one's behavior, and attitudes and beliefs about the likely outcomes of a behavior. Consistent with the theory, Parker and colleagues (1992) found that intentions to drive in a risky manner were associated with less concern with social norms regarding safe driving, higher perceptions of driving skill, and less worry about the consequences of unsafe driving. Although the evidence linking perceived norms and self-efficacy to risky driving is limited, the importance of these constructs in other health interventions suggests that attempts to change service members' risky driving by changing their beliefs will be far more effective if they focus on norms, control over behavior, and attitudes about outcomes, in addition to risk perception.

The specifics of how best to do this should be determined by formative research, for example, using focus groups of service members to identify the key themes that resonate with them and the media or community outlets that reach them. But an approach that targets these important beliefs might involve a television advertisement in which an admired service member models safe behavior (e.g., puts on a seat belt or decides not to drive after a couple of drinks), emphasizes that most service members drive safely and only a few cause problems, describes why safety is important to him (for example, safety promotes mission readiness or protects his family from harm), and discusses what might happen if he drove less safely (e.g., he might kill someone, or be unable to deploy because of injury in an accident).

Although it also has received little research attention, another set of beliefs that may hold promise for understanding and addressing risky driving in the military are those about masculinity. Identifying with the stereotypical male role is linked to driving behavior (Özkan and Lajunen, 2006), and identification with a "macho" personality has been specifically linked to aggressive driving (Krahé and Fenske, 2002). In a small but particularly well-conducted study examining this issue, Mast et al. (2008) tested whether males who were "primed" to think about masculinity changed their behavior. Participants were exposed to words related to masculinity or femininity, or to neutral words, and their behavior in a driving simulator was observed. Those that heard the masculine words increased their speed over a baseline assessment by approximately twice as much as did participants in the female and neutral word groups. As in any laboratory study, the relevance of these responses to driving behavior in a car might be questioned (for example, participants may have felt less reason to drive cautiously in a simulator than they would in a vehicle), but the study provides compelling evidence that thinking about masculinity causes greater risk-taking behind the wheel, as least among men, and at least in some situations. Together with the studies linking specific beliefs about masculinity to driving habits, this study suggests that decoupling any links that might exist in service members' minds between masculinity, bravery, and risky driving may be a fruitful avenue for reducing risky driving in this group.

Prior Drinking and Driving

Drunk drivers who were in fatal crashes have often driven drunk previously. NHTSA statistics showed that for drivers in fatal crashes with a BAC level of 0.08 or more, about 8 percent had previous convictions for DWI, and one-quarter had had their license suspended or revoked.[5] For drivers in fatal crashes who had not been drinking, these figures were one percent and 10 percent (NCSA, 2009a).

Examining fatal alcohol-related crashes in North Carolina over a ten-year period, Brewer et al. (1994) compared the records of drivers who died in these crashes and had a BAC level of 0.20 or higher with those of drivers who died in crashes but had lower BAC levels. Those with high BAC levels were more than eight times more likely to have been arrested in the past for driving while impaired. Even controlling for other variables, the study still found an odds ratio of 4.3 for drivers who had previously been arrested for DWI to be in a fatal crash within five years, compared with drivers who had not had an arrest for DWI. Each additional previous arrest added to prediction.

Prior Crashes

Another group to target is those who have already been involved in a crash. The likelihood of a crash in which the driver was at least partially responsible is much higher among drivers who had one or more such crashes in the preceding year (Elander, West, and French, 1993).

[5] The figure for drivers who had had a license suspended or revoked (the term varies by state) includes drivers whose licenses were suspended under a procedure called "administrative license revocation," under which suspected drunk drivers have their license revoked on the spot. This topic is discussed in more detail in Chapter Seven.

This relationship led West and colleagues to conclude that as much as 75 percent of the risk of motor vehicle crash involvement could be attributed to a stable characteristic of the driver—that some individuals are "crash prone." This relationship also exists among military personnel. Service members previously hospitalized because of a motor vehicle crash are more likely to die in a subsequent motor vehicle crash (Hooper et al., 2006). The relationship between prior and subsequent crashes may be based in driving skill or perceptual abilities, but it is likely that at least a portion of it is a consequence of habitual risky driving behavior. Hartos, Eitel, and Simons-Morton (2002) find that a composite "risky driving" measure shows stability over a three-month time period. This suggests that risky driving practices vary little across time—those who speed, run lights, etc. at one point are the same individuals most likely to do so in the same situation at a later time.

Deployment

Although evidence reviewed elsewhere in this report indicates that deployment is a risk factor for motor vehicle crashes among military personnel, evidence that it affects driving behavior is more limited. A recent British study found that overall, those deployed in the first phase of the current war in Iraq were somewhat more likely (adjusted odds ratio of 1.33) to drive in a risky manner than those who did not deploy, but this difference disappeared for the counter-insurgency phase. In looking at those who deployed, during both periods of the war the risk rose with increasing exposure to trauma, and for those in the counter-insurgency phase, the risks rose if the service member had problems at home. However, the length of deployment did not have an effect (Fear et al., 2008). A study of 1,250 U.S. soldiers recently returned from OIF in 2006 found that those who experienced violent combat or killed somebody scored somewhat higher on a risk-taking scale (measuring characteristics like impulsiveness and invincibility) than those who deployed but did not have those experiences. In terms of actual behavior, soldiers who experienced violent combat were more likely to drink heavily in one sitting, and those who had killed somebody (either an enemy or in a "friendly fire" incident) were more likely to behave aggressively (the survey did not ask specifically about driving behavior). The increases in behaviors were slight but statistically significant (Killgore et al., 2008). In a study comparing motor vehicle crash fatalities among Gulf War veterans to those among a group of non–Gulf War veterans (described in Kang et al., 2002), it was found that Gulf War veterans were more often driving in risky ways—they were not using seat belts or helmets, or they were speeding or drinking. The authors interpreted these data as suggesting that deployment dampens perceptions of risk, thereby leading to more risky driving behavior.

None of these studies was able to conclusively determine whether deployment itself was the main factor, or whether those who deploy are more likely to take risks to begin with. Bell et al. (2000a) found that active duty personnel who deployed in the Gulf War were more likely to have prewar hospitalizations for injury, and may also have engaged in more risky driving behaviors prior to the war (including driving after drinking, driving with a drinker, consistently speeding, and/or inconsistent seat belt use). Among men, those who deployed were also more likely to have received hazardous duty pay prior to the war, suggesting that they come from occupational groups within the Army that may involve riskier activity. While small sample sizes and limited information limit the conclusions that can be drawn, the results from this well-designed study suggest that, as a result of occupational choice within the military, those who deploy during wartime may have risk-taking attitudes that put them at greater risk for vehicle crashes when they return from deployment. Further, though indirect, support for this interpretation is provided by findings that military personnel who die in motor vehicle crashes come from occupational categories that may select for risk-taking personality (Hooper et al., 2006), and evidence from a meta-analysis that suggests that deployment may select for risk takers (Knapik et al., 2009).

CHAPTER SIX

Effectiveness of Safety Interventions

The military has taken a number of steps to reduce risky driving, particularly DWI (Moore and Ames, 2009). Education has been tried as an approach to reducing DWI, as well as campaigns to increase awareness. For example, the PREVENT program workshops on DWI attempt to educate service members about drinking and driving in order to reduce its incidence. The Air Force's current "0-0-1-3" campaign is meant to raise awareness of drinking and driving among service members by communicating that the Air Force has zero tolerance for underage drinking and DWI, and reminding personnel of the limit to the number of drinks they can safely consume before driving. Right Spirit is an awareness campaign by the Navy, and the Navy has also distributed "Don't Drink and Drive" stickers customized to include squadron names. The Navy has also developed designated driver and free ride programs. Bases sometimes have policies under which a specified number of days with no DWI incidents results in a day off or an event or special meal (with no alcohol) for all personnel. Unfortunately, these programs do not have a strong evidence or theoretical base, and we know of no formal evaluations of any of them. Thus, while we cannot speak to their effectiveness definitely, they cannot be described as good practice without such evaluation.

There is, however, a very well-developed literature testing interventions to reduce risky driving in the general population. Comprehensive reviews of a number of these approaches are provided in other reports (Sivak et al., 2006; NHTSA, 2007a; Transportation Research Board, 2009). Below, we briefly outline a number of these and describe their effectiveness, referring to military programs where relevant and possible.

Mandatory seat belt laws have the strongest record of reducing vehicle fatalities among the strategies reviewed. Of the drinking and driving policies we discuss, lowering legal limits on BAC levels for drivers, zero tolerance for underage driving, and random breath testing/sobriety check points are among the practices with the strongest research support and the greatest demonstrated effectiveness. Each of these measures has been the subject of multiple well-designed studies using either a comparison of alcohol-related accident rates prior to versus following implementation, a comparison of otherwise comparable geographic areas with and without the policy in place, or some combination of both research approaches. The best method of implementing these policies is through the practice of high-visibility enforcement (HVE), which involves not only enforcing rules, but widely publicizing this enforcement. Broader media campaigns have been labeled ineffective in some program reviews, but this is largely because many anti–drunk driving campaigns have been poorly implemented and unguided by theory and evidence-based practice. We conclude that when they are well-designed and implemented they do work. We also conclude that screening and intervention for alcohol use problems may be an effective strategy for the military. Though fewer studies have examined this approach to reducing drinking and driving, those that do indicate it is a promising practice in populations with a high prevalence of heavy drinking. We find that designated driver programs and technological approaches have the least evidence of effectiveness.

Media Campaigns

Media campaigns can be a highly effective method of changing risky behaviors in a broad population (Hornik, 2002). In a review of media campaigns to reduce DWI, Elder and colleagues (2004) estimate they reduce alcohol-related crashes by 13 percent. The decrease in injury crashes is slightly less—10 percent based on the studies in their review. Only carefully planned and executed campaigns were included in the analysis, so these estimates are likely among the highest observed in the research literature. Elder and colleagues' estimates may be conservative, nonetheless. Media campaigns are difficult to evaluate for a number of reasons (Hornik, 2002). Such campaigns are designed to reach mass audiences, and so may easily "spill over" into comparison groups. Changes in risky behavior over time that occur naturally as norms surrounding a behavior shift can obscure the effects of campaigns. For example, a general downturn in rates of motor vehicle crashes makes it more difficult to observe an additional reduction attributable to a media campaign (Elder et al., 2004). Finally, campaigns may induce indirect effects on behavior that go undetected in short term follow-ups. Yanovitzky and Bennett (1999) found that much of the effect of media coverage of drunk driving over a 19-year period was on legislation, and that legislation in turn influenced drunk driving behavior. They argue that the social and cultural changes in acceptability of risky driving wrought by media campaigns may be their most important outcome, and altered norms are critical in changing individual behavior. Thus, the real and long-term effects of a media campaign may not be evident using evaluation strategies currently available.

In general, media campaigns are most effective when they create awareness of an issue or inform people about things of which they were previously unaware (Hornik, 2002). Sivak et al. (2006) suggest that they work poorly for speeding, for example, because most people already view speeding as risky (though perceptions of one's own speeding as risky and of one's own speed as high tend to reflect less awareness; Svenson, 1981; Walton and Bathurst, 1998). However, campaigns targeting more complex beliefs have also been found to be effective. The Foolspeed campaign in Scotland, based on the Theory of Planned Behavior (Fishbein and Ajzen, 1975), is an excellent example. Foolspeed targeted perceptions of normative behavior (how fast do most people drive?), and perceived ability to prevent crashes with one's own behavior (self-efficacy), in addition to expected outcomes of speeding. The ads depicted realistic driving scenarios and pressures that often lead to speeding, showing drivers resisting these pressures as well as depicting the consequences of speeding (e.g., for vehicle control, getting to destinations quickly, and friends' and families' anxiety). Although there were no observed effects of the campaign on speeding behavior or perceived norms, comparison of baseline attitudes to those two years into the media campaign showed that it was effective in altering drivers' beliefs that speeding would make it difficult to stop quickly in an emergency, help them keep up with the flow of traffic, drive at a comfortable speed, and save time. Drivers exposed to the campaign saw fewer positive consequences and more negative consequences of speeding (Stead et al., 2005).

Randolph and Viswanath (2004) reviewed a broad set of media campaigns related to public health and identified five factors that contribute to their success. They include the following: (1) ensuring sufficient exposure to one's message, (2) using social marketing tools to create appropriate messages, (3) ensuring structural conditions create an environment supportive of change, (4) developing theory-based campaigns, and (5) including a process analysis that tests the effectiveness of campaigns during implementation (who are they reaching and with what reaction?). The latter allows for midcourse adjustment of the campaign and assists in interpretation of final results. In their review of campaigns designed to reduce alcohol impaired driving, Elder et al. (2004) reached similar conclusions. They also note that campaigns that fail to meet these criteria are seldom subjected to evaluations of sufficient rigor to allow assessment of their effects, so the conclusion that they are ineffective is largely based on anecdote.

A number of strategies may be employed to achieve the quality criteria these authors put forth. Exposure is often assured by the use of multiple media (print, broadcast, billboards and community events) over an extended time period (several months or more), and ensuring proper placement of messages within these media. Creating appropriate messages means including the targeted audience (risky-driving military personnel) in message development and testing, as well as employing professional communication services to produce and distribute the ads. Some work suggests that sensation-seekers—those most likely to be risky drivers and thus a key target of any driver safety campaign, may need targeted messages that tap their own interests and dispositions. Rosenbloom (2003) showed young men a videotaped message featuring a graveyard where each gravestone was marked with a driving speed (e.g., "80 mph") and found that high sensation-seekers drove faster after seeing the film while low sensation-seekers drove slower afterwards. This is consistent with some of the evidence concerning fear appeals discussed later in this chapter, but adds to it by suggesting that some of the riskiest drivers may be those most resistant to these forms of appeals.

In a similar vein, Ulleberg (2001) examined responses to a Norwegian driver safety campaign launched at schools across the country. The campaign involved two videos shown at schools, as well as follow-up in-school activities and supporting media messages. While most adolescents rated the campaign favorably, those who were high sensation-seekers and also had more aggressive personalities reported that they were uninterested in the message and rated it poorly. These individuals also had higher levels of traffic accidents and more accident-related injuries than other students, underscoring that they were the most important targets of such a campaign.

Campaigns Targeting Sensation-Seekers

In response to such findings, researchers have begun to develop health and safety campaigns specifically aimed at reaching sensation-seekers, an approach termed "SENTAR." SENTAR (sensation targeting) has been used successfully to reduce use of marijuana among adolescents by enlisting potential users (who tend to be sensation-seekers) in message development and targeting them with the resulting novel, fast-paced, and physically and emotionally arousing messages (Palmgreen et al., 2002).

Randolph and Viswanath's (2004) third point in the list above, ensuring that structural conditions will support change, might involve examining the regulations in place regarding seat belt use, drinking and driving and their enforcement, or the conditions under which alcohol is sold or consumed on and off base. Structural conditions might also involve driver safety training, access to counseling and treatment for stress or substance use, or the norms surrounding drinking and drunk drinking. The synergy between supportive environments and safety campaigns is perhaps best exemplified in the success of HVE campaigns, discussed later in this chapter.

Media Campaigns Using "Fear Appeals"

Many of the media campaigns that have been developed around the issue of traffic safety use an approach referred to in the psychological literature as a "fear appeal." Fear appeals involve efforts to convince a target audience that they are threatened by a negative outcome—that it is both likely to befall them and involves substantial negative consequences. Multiple meta-analyses and reviews driven by various theoretical perspectives have been conducted examining the effects of fear-inducing messages, including traffic safety appeals. For example, a review of campaigns intended to reduce drinking and driving (Cismaru, Lavack, and Markewich, 2009) concluded that fear campaigns effectively did so. Reviews of the more general literature examining fear appeals reach similar conclusions. They have found that fear arousal can be an effective way of changing attitudes and motivating behavior, though effects observed have been small. Correlations of level of fear with attitudes, intentions to change behavior, and behavior are in the 0.13 range, implying a weak relationship (Witte and Allen, 2000).

However, communication designed to arouse fear can backfire. This is because one can control fear by changing behavior (e.g., driving more safely), or by reacting defensively or avoiding the threatening information (dismissing the message or trying to prove it wrong). In their meta-analysis, Witte and Allen (2000) found that messages that provoked higher levels of fear, but left recipients feeling unable to change their behavior in such a way as to avoid the threat (i.e., feeling low in "response efficacy"), resulted in attempts to control emotional reactions. These in turn resulted in less persuasion, as people tried not to think about the message or see it as self-relevant. Similarly, in a review of studies testing the theory underpinning the use of fear appeals, protection motivation theory, the authors concluded that while both feelings of fear and those of response efficacy play a role in behavior, response efficacy is the more influential factor (Milne, Sheeran, and Orbell, 2000).

A second, related reason to avoid traffic safety messages designed to induce fear is the literature on message framing. Studies in this area suggest that public health messages that focus on what people have to lose by failing to comply with recommendations are useful in promoting screening and detection of disease; messages that focus on what could be gained by compliance are more effective for inducing preventive health behavior (Rothman and Salovey, 1997). Thus, focusing on the risks of drunk driving or speeding may be less effective than messages linking safe driving with military readiness or the safety of family and friends, as long as the message remains compelling and relevant to its recipient. In the traffic safety literature in particular, Sibley and Harré (2009) found that positively framed messages, showing an individual engaging in safe driving, were less likely to provoke a defensive increase in estimates of one's driving ability than were negatively framed messages. Because they are framed in terms of potential loss, fear appeals may provoke defensive shifts in beliefs that are counterproductive to changing risky driving behavior.

Messages in media or training venues can promote response efficacy by focusing on how specific safe behaviors reduce crashes and deaths. For example, noting that sobriety checkpoints reduce the number of vehicle crashes in an area, or that wearing a seat belt is the single most important factor in determining whether a person survives a crash.

If fear appeals are used, it is important to keep a few caveats in mind. First, based on the literature just described, it is wise to pair fear appeals with the portrayal of behaviors that substantially reduce risk, as well as with information designed to build message recipients' confidence in their ability to engage in these behaviors. Second, fear-inducing messages do not need to be graphic or gruesome. Evidence suggests that information presented less vividly will be equally effective (de Hoog, Stroebe, and de Wit, 2007), and narratives will be more persuasive than statistical descriptions of risk (de Wit, Das, and Vet, 2008). Finally, some studies suggest that it is more important to communicate to people that they are at risk, and only secondarily, that the risk they face is extreme (de Hoog, Stroebe, and de Wit, 2007).

High-Visibility Enforcement

High-visibility enforcement involves implementing new enforcement measures for existing regulations (e.g. a primary seat belt law or BAC level checkpoints) along with campaigns to make the public aware of these measures, and typically, to also persuade the public of the advantages of complying with these regulations. Overall, HVE campaigns have been highly successful.

Most HVE programs are based on the Canadian model, Special Traffic Enforcement Program, which produced very high rates of seat belt use. The Canadian experience makes clear the key principle of HVE: Laws work better when they are enforced, and this enforcement is widely publicized. When provincial laws were first implemented requiring that seat belts be worn, use rates remained low (in the 40–60 percent range). But when governments began enforcing these laws, publicizing this, and also publicizing the important role of seat belts in safety, rates of use quickly escalated, with these combined efforts believed to be responsible for current use of seat belts by about 90 percent of the population (Dussault, 1990; Landry, 1991).

In the United States, examples of HVE programs are the click it or ticket (CIOT) campaigns to increase seat belt use, the use of DWI checkpoints, and the Ticketing Aggressive Cars and Trucks program used to reduce aggressive driving in the state of Washington. CIOT was implemented in Washington State in conjunction with a shift to a primary seat belt law.[1] The effort improved seat belt use from 83 percent to 93 percent after one year and 95 percent after two years (Salzberg and Moffat, 2004). The 2003 National CIOT program was also highly effective, increasing seat belt use in 40 of 47 states (Solomon, Chaudhary, and Cosgrove, 2003), and the 2005 program showed increases in 35 of 47 states (Solomon et al., 2007). Seat belt use on bases is currently mandatory for service members but it may be possible, as in the Canadian experience, to enhance use through more rigorous and well-publicized enforcement of this regulation.

Programs similar in concept to HVE have been implemented fairly successfully on military bases, although they were carried out before seat belt use was mandatory. One study compared two U.S. Navy bases that required staff to wear seat belts at the time. Both bases implemented short-term programs to encourage seat belt use: One base ticketed sailors who were not wearing seat belts for three weeks, and the other entered sailors who were wearing seat belts into a prize drawing for four weeks. At the base issuing tickets, seat belt use rose

[1] State seat belt laws are either primary or secondary. Primary laws allow police to pull drivers over and ticket them for failure to wear a seat belt. Secondary laws allow police to ticket drivers for failing to wear a seat belt, but only if they have been ticketed for another offense.

from 55 to 79 percent, while at the base with the prize drawing, seat belt use increased from 51 to 61 percent. Men and women responded equally to both programs, but six months after the campaigns ended, women were significantly more likely to keep wearing their seat belts, while men's rates declined (Kalsher et al., 1989). In the Netherlands, similar enforcement and incentive campaigns were carried out on 12 bases. Both types of campaigns increased seat belt use (67 to 75 percent for enforcement programs, and 62 to 69 percent for incentive programs), and those campaigns with extra publicity did not achieve any greater increases. The greater the police effort in enforcement, as measured in number of fines per hour, the higher the increase in seat belt use. The greatest increases in seat belt use were for personnel under 25. Seat belt use rates remained at post-campaign levels even three months later (Hagenzieker, 1991).

In a model similar to the CIOT program, sobriety checkpoints have been effective in reducing DWI when paired with sufficient publicity regarding enforcement of DWI laws (and when the number of checkpoints is sufficient to lead potential intoxicated drivers to expect they might be caught). An evaluation of such programs in 7 states found reductions of up to 20 percent in alcohol-related fatalities (Fell et al., 2008).

Because failure to use seat belts and drinking and driving often co-occur (Nichols et al., 2009), it is also possible to create combined HVE programs that target both. A good example of this is the "Buckle Up and Drive Sober" campaign implemented in Binghamton, NY. The two-year effort reduced drinking and driving, late night and injury crashes, and increased seat belt use, particularly at night (Wells, Preusser, and Williams, 1992).

It is important to note that not all HVE campaigns have succeeded. For example, programs targeting aggressive driving in Marion County, Indiana, and Tucson, Arizona were associated with a decrease in the proportion of crashes due to aggressive driving in Tucson, but an increase in Marion County (Stuster, 2004). The program evaluator suggested that this might have been the result of a greater emphasis on enforcement in Tucson, or to the concentration of enforcement duties among a small set of Tucson's officers. Two officers were assigned full time to the task and two part time, while in Marion County enforcement duties were distributed broadly across officers and agencies. Nonetheless, both programs were well-planned and executed, and engaged the processes believed to underlie successful HVE campaigns.

Changes in Regulation or Enforcement

Administrative license revocation or suspension is a highly effective strategy used in the general population to address drinking and driving. These regulations allow arresting officers or other officials to remove a driver's license for a fixed period of time, before any court proceedings have taken place, if a BAC test is failed or refused. The driver can request a hearing, and the arrest may or may not lead to a criminal conviction (NHTSA, 2008a). The combination of swift and certain punishment is thought to drive the success of this measure, which has been estimated to reduce alcohol-related fatal crashes by 5 percent. (Wagenaar and Maldonado-Molina, 2007). Administrative license revocation or suspension is currently in place in 41 states, and 32 states require some drivers with multiple offenses to forfeit their vehicles.[2]

[2] As of December 2009, nine states do not have administrative license suspension laws: Kentucky, Michigan, Montana, New Jersey, New York, Pennsylvania, Rhode Island, South Dakota, and Tennessee.

Server trainings for bartenders appear to be effective at reducing drinking and driving (Shults et al., 2001), especially when they are backed by a strong management commitment to responsible beverage service and enforcement of laws against selling to intoxicated persons, but more research is needed (Wagenaar and Tobler, 2007). A responsible beverage service intervention in a Navy bar was found to lower rates of intoxication and DWI (Saltz, 1987).

Evidence concerning the effectiveness of ride programs for reducing drinking and driving is mixed, and there is no evidence of success for designated driver programs, which tend to be ill-defined and have not been systematically evaluated (Grube, 2007).

The increase in the U.S. of the minimum legal drinking age from 18 to 21 (Hedlund, Ulmer, and Preusser, 2001; O'Malley and Wagenaar, 1991) and adoption of zero tolerance laws that require BAC levels of 0.02 or less among drivers under age 21 (Hingson, Heeren, and Winter, 1994; Voas, Tippetts, and Fell, 2003) have been found to reduce DWI among youth under 21, and the minimum legal drinking age effect may persist beyond age 21 (O'Malley and Wagenaar, 1991). Zero tolerance has also reduced alcohol-related crashes (Hingson, Heeren, and Winter, 1994; Shults et al., 2001). These regulations appear to operate by reducing overall drinking, the perceived acceptability of drinking and driving, and drunk driving among youth. Although it may not be practical for the military to institute its own BAC level standards or age limits for drinking, it may be possible to more rigorously enforce existing laws among service members. As noted in Chapter Five, a large portion of drinking and driving among service personnel involves underage drinkers.

Treatment and Prevention of Alcohol Problems

Screening and treatment for alcohol problems can be a useful method of reducing alcohol-related crashes (Wells-Parker et al., 1995), as can intervention with individuals at high risk for alcohol problems who may not meet criteria for an alcohol dependence diagnosis (Dill, Wells-Parker, and Soderstrom, 2004). Although alcohol treatment can be costly and time intensive, brief interventions can be delivered by trained nonexperts, in as little as one session, or as part of a series of small group discussions (e.g., D'Amico and Edelen, 2007; Mundt et al., 2005).

An additional benefit of these approaches is that alcohol treatment/prevention may help to uncover and address other mental health problems that appear to exacerbate drinking and which are present at very high rates among DWI offenders. In a sample of individuals convicted of DWI in the Western United States, 30 percent of women and 36 percent of men had a diagnosis of alcohol dependence in the past 12 months. Among those, 50 percent of women and 30 percent of men had a co-occurring psychiatric condition during the same time period, most commonly major depression or PTSD. The overall rate of past year PTSD in the sample was 16.5 percent for women and 6.5 percent for men; rates of major depression were 17 percent for women and 6.6 percent for men (Lapham et al., 2001). A recent analysis of repeat DWI offenders in the Eastern United States (with two or more DWI convictions) found that nearly three-quarters of these individuals (73.5 percent) had past year alcohol use disorders. In this primarily male sample (81 percent), past year PTSD occurred in 11.5 percent of the sample and past year major depression in 8.2 percent (Shaffer et al., 2007). Both studies found that rates of PTSD and depression in the past year were significantly higher among those with a DWI conviction than those in the general population, even after adjusting for demographic differences.

Rates of PTSD and depression among military personnel are not known, but questions included in the most recent HRBS (Bray et al., 2009), which indicate a probable diagnosis and the need for further evaluation, identified 21 percent of personnel as possibly depressed, and 11 percent as possibly having PTSD. Current probable PTSD and depression rates are 14 percent among previously deployed individuals in the military (Tanielian and Jaycox, 2008). In the U.S. general population, the estimated rate of major depression is 6.6 percent (Kessler et al., 2003) and for PTSD it is 3.5 percent (Kessler et al., 2005). Rates of these disorders in the general population are not demographically adjusted to indicate clearly how the military compares, but the rates of probable disorder among ADMP, together with the aforementioned data linking both PTSD and depression with DWI, indicate it is likely that high levels of both disorders are present among military DWI offenders. Treating these disorders has inherent value for improving military health, and to the extent that the disorders promote drinking, treatment is likely to also reduce rates of DWI among military personnel.

Technological Approaches

Some have proposed that making technology such as breath analysis testing available to individuals will allow them to monitor their BAC levels and avoid driving when they have had too much to drink. The Army has distributed to a number of troops key chain breath analysis testers, customized with a slogan chosen by each commander, and these have met with positive responses (Reister-Hartsell and Gardiner, 2008). However, emerging evidence suggests that there may be unintended negative consequences of making breath analysis testing or other consumption-tracking devices readily available to drinkers. In one of a recent series of studies, researchers gave drivers an incentive to keep their BAC levels under 0.05 and supplied some of them with know-your-limit matrices to allow them to estimate their own BAC levels. Those who were given the matrices actually did worse at staying under the 0.05 limit than drivers simply warned not to drink and drive (Johnson, 2009). The author speculated that the ability to measure one's BAC level may provide a tool for drinkers to keep their consumption as high as possible while remaining under the legal limit. If so, and this finding generalizes to personal breath analysis testers, then these devices could actually lead to greater numbers of impaired drivers on the road.

CHAPTER SEVEN

Effectiveness of Motorcycle Training and Other Policies

This chapter looks at some of the ways that have been tried to reduce the number of motorcycle crashes and their lethality, and how effective they have been. Universal helmet laws are by far the most important policy in this regard, as they have been shown over and over to be effective in increasing helmet use and saving the lives of motorcyclists in crashes.

The effectiveness of other motorcycle safety policies is not as certain. Studies of licensing restrictions, such as learner's permits, graduated driver licensing (GDL), and tiered licensing, have found mixed results. Although rider training programs are widespread and supported by riders, riding groups, the motorcycle industry and NHTSA, evidence is mixed as to whether training reduces motorcycle crashes. Some studies even found that motorcycle rider training increased the likelihood that a rider would be involved in a crash. While helmets have been shown to prevent fatal injuries to motorcyclists, other types of personal protective equipment (PPE) have only been shown to be effective in preventing soft tissue injuries, but not fractures, serious injuries, or fatalities during crashes. These mixed results indicate that the effectiveness of safety policies, such as licensing restrictions, rider training, and protective equipment, on reducing motorcycle fatalities requires more research.

Helmet Laws

As described in Chapter Two, helmets prevent traumatic brain injuries and fatal injuries to motorcyclists. The NHTSA 2006 Motorcycle Safety Program Plan states that "helmet use laws are the most effective way to get all motorcyclists to wear helmets" (NHTSA, 2006a). As of July 2010, 20 states and the District of Columbia had universal helmet laws that required helmet use by all motorcyclists regardless of age or experience.[1] Twenty-seven other states have partial helmet laws that required only motorcyclists under a certain age (usually 18) to wear helmets, and three states (Illinois, Iowa, and New Hampshire) had no laws requiring helmet use (IIHS, 2010a). In 2009, legislation was introduced in 19 of the 20 states that had laws requiring all riders to wear helmets, attempting to repeal the law; however, none of the bills was passed. Since 1997, six states have repealed their universal helmet laws, enacting instead partial

[1] The 20 states are Alabama, California, Georgia, Louisiana, Maryland, Massachusetts, Michigan, Mississippi, Missouri, Nebraska, Nevada, New Jersey, New York, North Carolina, Oregon, Tennessee, Vermont, Virginia, Washington, and West Virginia (IIHS, 2010a).

helmet laws that limit coverage to riders under a specific age, generally 21 (NHTSA, 2008c; Houston and Richardson, 2007).[2]

According to IIHS, "the best way to reverse the trend of declining helmet use and rising deaths is to require all motorcyclists to wear helmets" by enacting universal helmet laws (IIHS, 2007, p. 4). An extensive review of the literature found that universal helmet laws are significantly associated with an increase in helmet use and decreases in motorcycle fatalities, head injuries, days of hospitalization, and medical costs (Lin and Kraus, 2009). In contrast, partial helmet laws that limit coverage to riders under a specific age are not effective at protecting young motorcyclists. Data show that that fewer than 40 percent of fatally injured young motorcyclists were wearing helmets even though they are legally required to do so, and helmet use will likely continue to decline because laws that only govern young riders are difficult to enforce (Muller, 2004; Kyrychenko and McCartt, 2006; NHTSA, 2008c).

Studies have found that helmet laws increase helmet use and reduce motorcycle fatalities, based on both comparisons across states and before-and-after assessments of enacting or repealing helmet laws. For example, a U.S. study comparing motorcycle mortality rates between states with and without helmet laws from 1997 to 1999 found an approximately 20 percent lower mortality rate in states with full helmet laws. Other studies have found similar results (McGwin et al., 2004). Data from all 50 states and the District of Columbia from 1975 through 2004 showed that states with universal helmet laws had an 11 percent reduction in motorcyclist fatalities as compared to states with no helmet laws, while states with partial coverage laws were not statistically different from states with no helmet law (Houston and Richardson, 2007). The first year after universal helmet laws were enacted in five states, motorcycle fatalities decreased by: 33 percent in Oregon; 32 percent in Nebraska; 15 percent in Washington; 37 percent in California; and 20 percent in Maryland (NHTSA, 2008c).

When states changed their helmet laws from universal to partial, helmet use decreased and fatality rates increased (Preusser et al., 2000; Houston and Richardson, 2007; NHTSA, 2008c):

- In Arkansas and Texas, helmet use dropped from 97 percent under the universal helmet law to 52 percent in Arkansas and 66 percent in Texas within nine months of the change in law (Preusser et al., 2000). Helmet use decreased by approximately 10 percent in young injured motorcycle riders in Texas, who were still required by law to wear helmets under the new partial helmet law (Preusser et al., 2000). In Kentucky and Louisiana, helmet use dropped from nearly full compliance under the universal law to approximately 50 percent with the partial law (NHTSA, 2008c).
- In the first full year following repeal of the universal helmet laws, motorcyclist fatalities increased by 21 percent in Arkansas and 31 percent in Texas (NHTSA, 2008c). Motorcyclist fatalities increased by over 50 percent in Kentucky and over 100 percent in Louisiana (NHTSA, 2008c). Injuries also increased 48 percent in Louisiana and 34 percent in Kentucky (NHTSA, 2008c). Looking at motorcyclist fatalities in the three full years after the laws changed shows that fatalities increased by 23 percent in Arkansas, 52 percent

[2] Arkansas and Texas were the first to change their helmet laws in 1997, followed by Kentucky in 1998, Louisiana in 1999, Florida in 2000, and Pennsylvania in 2003 (Preusser, Hedlund, and Ulmer, 2000; Houston and Richardson, 2007). Louisiana has since reenacted a universal helmet law.

in Texas, 99 percent in Kentucky, and 130 percent in Louisiana (Ulmer and Northrup, 2005).
- Since 1997, an estimated 615 additional motorcyclist fatalities occurred in the 6 states that repealed their universal helmet laws, a 15 percent increase over what would have been expected had universal coverage not been repealed (Houston and Richardson, 2007).

Several studies have examined the effects of Florida's change from a universal to a partial helmet law and found substantial decreases in helmet use and increases in injuries and fatalities (Muller, 2004; Ulmer and Northrup, 2005; Kyrychenko and McCartt, 2006).[3] Florida is of interest because it has the highest per capita motorcycle fatality rate and accounted for approximately 11 percent of all motorcycle fatalities in the United States in 2008 (see Chapter Two):

- In Florida, helmet use decreased from 65 percent under the universal helmet law to 47 percent under the partial helmet law (Ulmer and Northrup, 2005). Helmet use declined markedly among riders under the age of 21, even though they were still required to wear a helmet under the amended law (Ulmer and Northrup, 2005).
- Motorcyclist death rates increased for single- and multiple-vehicle crashes, for male and female operators, and for riders of all ages including those younger than 21 (Kyrychenko and McCartt, 2006). After controlling for gender and age, the likelihood of dying in a motorcycle crash was 25 percent higher than expected after the law change (Kyrychenko and McCartt, 2006).
- Motorcycle registrations and VMT increased after Florida repealed its universal helmet law (Muller, 2004; Ulmer and Northrup, 2005). Controlling for VMT, motorcyclist fatalities increased by 38 percent (Muller, 2004). Controlling for motorcycle registration, fatalities increased by 21 percent per 10,000 registered motorcycles two years after the law change (for the United States as a whole, fatalities increased by 13 percent nationally over the same period) (Ulmer and Northrup, 2005).
- Deaths of riders not wearing helmets and under age 21 increased 188 percent, even though they were still required to wear a helmet by the amended law (Ulmer and Northrup, 2005).
- It is estimated that between 46 and 82 motorcyclists died in Florida the first year after the universal helmet law was repealed, and 117 fatalities could have been avoided during 2001–2002 if Florida's universal helmet law had remained in place (Muller, 2004; Kyrychenko and McCartt, 2006).

Licensing Restrictions

To obtain a motorcycle operator license, motorcyclists must demonstrate knowledge of the rules of the road, riding skills, and the ability to safely operate a motorcycle (Baer, Cook, and Baldi, 2005). Motorcycle rider licensing restrictions are designed to allow novice motorcyclists to gain some experience before achieving full license status and to minimize exposure to risky situations (Reeder et al., 1999). Three basic types of licensing restrictions are commonly used

[3] On July 1, 2000, Florida's universal helmet law was amended to exempt riders 21 years of age or older from wearing helmets provided they have at least $10,000 of medical insurance coverage.

for motorcyclists: learner's permits, GDL, and tiered licensing. All states require a motorcycle license or endorsement for on-street operation of a motorcycle, but licensing practices and the use of various licensing restrictions vary from state to state (Baer, Cook, and Baldi, 2005).

A learner's permit is a restricted license issued to novice riders and is required for on-street operation of a motorcycle. A learner's permit allows novice riders to practice riding skills and to minimize exposure to risky situations before being granted a full motorcycle operator license. In 2008, 48 states and the District of Columbia required a learner's permit for new motorcycle riders. The minimum age requirement for applying for a learner's permit ranges from 14 to 17 years old. Learner's permits are valid for varying lengths of time, ranging from 30 days to four years, and usually carry various riding restrictions, such as requiring supervision, no freeway driving, no night time driving, no riding with passengers, and mandatory helmet and eye protection use. In addition, all states issuing learners' permits required applicants to pass a knowledge test about the rules of the road, but only six states also required applicants to pass a skills test to demonstrate safe operation of a motorcycle (MSF, 2008).

The purpose of a GDL system for motorcycle riders is similar to that of the learner's permit. However, the GDL usually has three stages—a learner's permit, an intermediate (sometimes called provisional) license, and a full license—that require a period of supervised instruction and place restrictions on the novice driver (Hanchulak and Robinson, 2009).

GDL systems, which originated in New Zealand, have been in use for over 20 years for automobile drivers (Reeder et al., 1999). In 2009, 49 states and the District of Columbia had three-stage GDLs for automobile licensing (IIHS, 2010b). In contrast, only 15 states were using GDLs for motorcycle licensing in 2008 (Hanchulak and Robinson, 2009). Some states do not have GDL programs specifically for motorcyclists, but require riders under the age of 18 to abide by the GDL provisions in place for automobile drivers (NHTSA, 2005b).

NHTSA developed a model GDL system for new motorcycle riders (Hanchulak and Robinson, 2009):

- A learner's permit is required for on-street operation by all applicants who are under 21, or any rider who does not hold a valid operator's license or permit at the time of application. The learner's permit provides novice riders with a minimum period of time to practice and develop basic motorcycle skills before riders progress to the intermediate license. A learner's permit allows driving only while supervised by a fully licensed driver, such as a parent or driving instructor.
- An intermediate license provides novice riders with additional time to gain practical experience and develop skills and behaviors associated with safe riding. The intermediate license still imposes several restrictions but allows unsupervised driving under certain restrictions.
- A full or unrestricted license is the third stage.

NHTSA "strongly supports" graduated licensing, but acknowledges that two elements associated with it, supervised operation and parental involvement, are problematic with motorcycles (NHTSA and MSF, 2000). One suggested solution to these problems is having a fully licensed rider at least 21 years old with at least three years of experience always be on another motorcycle accompanying the permit holder (Hanchulak and Robinson, 2009).

A tiered licensing system restricts a rider's operation of a motorcycle based upon its engine size and sometimes the rider's age (NHTSA and MSF, 2000). The United Kingdom, Australia, and New Zealand have tiered licensing for motorcyclists based on engine size and riding experience. Tiered licensing has not been widely implemented in the United States (NHTSA and MSF, 2000). In 2008, only nine states had tiered licensing for motorcyclists. Generally these systems restrict young riders (e.g., usually under the age of 18) to riding smaller bikes (typically between 100–250 cc in size) (Hanchulak and Robinson, 2009).

Studies of these various licensing systems have found mixed results. An evaluation of the association between motorcycle licensing and mortality rates in the United States from 1997 through 1999 found lower mortality rates in states that required skill tests for obtaining a learner's permit, had longer durations for learner's permits, or placed three or more restrictions on the learner's permit. However, tiered licensing and GDLs were not associated with lower mortality rates (McGwin et al., 2004). Other studies on tiered licensing have not shown reductions in crashes (NHTSA and MSF, 2000). A study in New Zealand found that the introduction of the GDL system for motorcyclists was closely related to a 22 percent reduction in hospitalizations due to motorcycle crashes for 15- to 19-year-olds (Reeder et al., 1999). The study concluded that the reduction in injuries was mainly attributable to an overall reduction in exposure to motorcycle riding (Reeder et al., 1999). A study of GDLs for motorcyclists in the United States had similar findings (Baldi, Baer, and Cook, 2005). Based on the mixed findings of studies on GDLs and tiered licensing, the effects of the GDL and tiered licensing systems on motorcycle fatalities require more research.

Rider Training Programs

The National Agenda for Motorcycle Safety describes motorcycle rider training and education as "the centerpiece of a comprehensive motorcycle safety program" (NHTSA and MSF, 2000). In 2006, 47 states had state-operated and legislated rider education and training programs (NHTSA and MSF, 2006).[4] In addition, many manufacturers offer rider training courses through their dealerships, such as Harley-Davidson and American Honda (IIHS, 2007; NHTSA and MSF, 2006). Rider training is required by some states for all motorcycle riders under a certain age (i.e., 16, 18, or 21 years of age, depending on the state), and in July 2008, Florida began requiring training for first-time applicants of all ages for motorcycle licenses. Many states waive some license test requirements, such as the knowledge or skills test, for motorcycle riders who have successfully completed an approved training course. Rider education and training programs are widely supported by riders, riding groups, the motorcycle industry, and NHTSA (NHTSA and MSF, 2006).

The basic *RiderCourse* and the experienced *RiderCourse Suite* curricula developed by MSF are used by training programs in most states (NHTSA and MSF, 2006). Approximately 4.5 million riders have graduated from the MSF rider training courses since 1974, and the number of students trained in MSF courses almost tripled over the ten-year period from 1996 to 2006 (Morris, 2009).

The military requires all of its riders to take a motorcycle training course that has been approved by either MSF or the Office of the Deputy Under Secretary of Defense (Installations and Environment). In addition, the Navy requires all sailors who ride sport bikes to take

[4] Alaska, Arkansas, the District of Columbia, and Mississippi are the only places that do not require rider education and training (NHTSA and MSF, 2006).

the MSF Military Sport Bike Course. The MSF Military Sport Bike Course is also offered to soldiers at a several Army bases. The Air Force's Air Mobility Command developed a mandatory sport bike safety class for all its airmen. The 2nd Marine Expeditionary Force at Camp Lejeune, N.C., which leads the Marine Corps' sport bike safety effort, is contracting with a professional motorcycle school and expanding the training to several Marine Corps sites (Miles, 2008).

Evidence is mixed as to whether rider training reduces motorcycle crashes. Studies have found a positive impact, no impact, or even a negative impact—that is, riders crashed more often after training. Studies that found that rider training did not reduce accidents include the following:

- In 1997, the Federation of European Motorcyclists reviewed 16 academic research papers conducted from 1979 to 1996 looking at the relationship between rider training and rider accidents. Eight of the papers reviewed concluded that the training did not reduce the likelihood of the rider being involved in an accident, seven said that rider training was effective, and one came to no conclusion either way (Federation of European Motorcyclists, 1997).
- A 1999 review of nine studies conducted from 1980 to 1995 looking at training using the MSF curriculum found that, when adjusted for gender, miles traveled, frequency of riding, and risk-taking behavior, the MSF course did not reduce the incidence of motorcycle crashes (Haworth, Smith, and Kowadlo, 2000). Of the nine studies, one found that untrained riders had 10 percent more crashes in the first six months of riding, while several other studies found that riders trained with MSF curriculum were more likely to have crashes and traffic offenses (Haworth, Smith, and Kowadlo, 2000).
- A study of Indiana motorcyclists in 2005 examining the effectiveness of rider training programs in reducing motorcycle accidents found that riders who took beginner rider training were more likely to be involved in an accident than those who did not and that riders who took the beginner training more than once were much more likely to be involved in an accident (Savolainen and Mannering, 2007).

Studies that found rider training to be effective in reducing accidents include the following:

- State-level longitudinal data for the continental United States from 1990 to 2005 was analyzed to determine whether alcohol and traffic safety policies, such as BAC limits, universal helmet laws, speed limits on rural interstates, and mandatory rider education, had an impact on motorcycle safety. State legislated or sponsored rider education that was mandatory for all or some riders reduced nonfatal injuries by approximately 10 percent, while universal helmet laws had the most significant effect on both nonfatal and fatal injuries. (French, Gumus, and Homer, 2009)
- An evaluation of the association between motorcycle licensing and operation regulations and motorcycle mortality rates in the United States from 1997 through 1999 found lower mortality rates in states that had driver training programs, skill tests for obtaining a motorcycle permit, longer durations for learner's permits, and full helmet laws (McGwin et al., 2004).
- An evaluation of trends in motorcycle accidents in California before and after the implementation of the California Motorcyclist Safety Program in 1987 found that fatal motor-

> cycle accidents decreased 69 percent and nonfatal accidents decreased 67 percent over the first nine years after the introduction of the California Motorcyclist Safety Program.[5] In particular, novice riders with less than 500 miles of riding experience prior to training had an accident rate less than half of those not undergoing training for at least six months after training. After six months, riding experience appeared to have a leveling effect on the differences between trained and untrained riders. In contrast, riders with more than 500 miles of riding experience prior to training had no significant differences in accident rates before or after taking the basic training course. In addition, a survey of trained and untrained riders in California found that riders who had undergone rider training were more likely to wear helmets and other protective gear (Billheimer, 1998).

Suggested explanations for these mixed findings include ineffective course material, changes in risk perception as a result of training, higher skill levels in riders who take training than those who do not, the difficulty of distinguishing between riders who want to be safer and the effects of training, and the differences between those who take training courses voluntarily and those who take training because it is mandatory (Savolainen and Mannering, 2007; Lin and Kraus, 2009).

The lack of evidence supporting the effectiveness of motorcycle training in reducing motorcycle crashes has been widely recognized. After a review of the literature, Lin and Kraus concluded that "the effectiveness of rider's education and training programs needs to be vigorously examined using better research designs (e.g., randomized controlled studies) and effective program components need to be identified" (Lin and Kraus, 2009, p. 719). Findings from another study indicate the need for a careful and comprehensive study of rider skills and risk perceptions to maximize the effectiveness of motorcycle training courses (Savolainen and Mannering, 2007). In the National Agenda for Motorcycle Safety, NHTSA also recommended research into the effectiveness and impact of rider education and training, and establishing benchmarks for rider education and training effectiveness (NHTSA and MSF, 2000).

NHTSA, MSF, and the National Association of State Motorcycle Safety Administrators have provided strategies and recommendations to improve, expand, and evaluate motorcycle rider training (NHTSA and MSF, 2006). In the National Agenda for Motorcycle Safety Implementation Guide, NHTSA identified important subjects that should be included in training and education courses to address key safety issues: information on roadway and other vehicle hazards; safe riding practices including braking, lane use, and defensive riding strategies; the dangers of alcohol and other drugs; the importance of wearing DOT-compliant helmets and other protective equipment; and strategies for the conspicuity of motorcycles and their riders (NHTSA and MSF, 2006). The Implementation Guide also recommended that the quality of training be monitored regularly both through student evaluations and independent reviews (NHTSA and MSF, 2006). NHTSA also developed a "promising practices" framework for rider education and licensing based on the key features of high-quality training identified in the National Agenda for Motorcycle Safety (NHTSA, 2005b).

---

[5] The California Motorcyclist Safety Program is a statewide program that is mandatory for riders under the age of 21 to obtain a California motorcycle license (Billheimer, 1998).

Personal Protective Equipment

Motorcycle protective apparel can (1) prevent or minimize injuries in a crash; (2) protect the rider from environmental conditions, such as wind, rain, and temperature; and (3) increase rider conspicuity (de Rome and Stanford, 2003). Because motorcyclists are usually separated from their motorcycles during a crash, protective apparel is more likely to protect the rider than protective equipment attached to the motorcycle (NHTSA and MSF, 2000). Injuries to the leg, such as fractures and soft tissue injuries, are the most common injuries in motorcycle crashes, affecting 30 to 70 percent of injured riders (Hurt, Ouellet, and Thom, 1981; Otte, Schroeder, and Richter, 2002; de Rome and Stanford, 2003; Lin and Kraus, 2009). Protective apparel can include a DOT-compliant helmet (see Chapter Four), eye protection, gloves, boots, and full coverage clothing such as pants and a jacket, or a one piece suit (NHTSA and MSF, 2000). Pants, jackets, and one piece suits are typically made of leather, Kevlar, or other fabric with high abrasion and tear resistance, and usually also include impact protectors such as padding, hard-shell material, or other impact-absorbing material fitted to the protective apparel to absorb or distribute force at specific impact points, such as the elbow, knee, spine, and shoulder (NHTSA and MSF, 2000; de Rome and Stanford, 2003). Gloves are typically made of leather or Kevlar and some have carbon fiber knuckle protection. Boots usually have reinforced ankles and toes.

Except for helmets, the United States does not have standards governing PPE for motorcyclists (NHTSA and MSF, 2000). In Europe, the European PPE Directive sets standards for all PPE, including motorcycles.[6] The European standards cover boots, gloves, clothing, goggles, and elbow, shoulder, knee, and spinal armor (Motor Cycle Council of NSW, Inc., 2009). In 2006, 36 states required motorcyclists to wear some form of eye protection, such as helmets with face shields, sun glasses, or goggles (NHTSA and MSF, 2006). Eye protection can help reduce vision impairment caused by wind and road debris (Hurt, Ouellet, and Thom, 1981; Otte, Schroeder, and Richter, 2002).

Several studies, including a comprehensive review of research on motorcycle protective clothing (de Rome and Stanford, 2003), have found that protective clothing reduces the frequency and extent of soft tissue injuries, such as abrasions and lacerations of the skin and soft tissue, exhaust and friction burns, and the stripping away of skin and muscle (Hurt, Ouellet, and Thom, 1981; Otte, Schroeder, and Richter, 2002; de Rome and Stanford, 2003). Heavy boots and work shoes can protect against ankle and foot injuries (Hurt, Ouellet, and Thom, 1981; Otte, Schroeder, and Richter, 2002). There is also evidence that well-designed protective clothing can contribute to personal comfort and reduce the risk of crashes due to fatigue, discomfort, and distraction caused by wind, rain, heat, and cold (de Rome and Stanford, 2003). In addition, it has been shown that in a crash, wearing protective clothing reduces the cost of motorcycle injury (Lawrence, Max, and Miller, 2002). However, protective clothing is not effective in preventing fractures or other serious injuries in high-impact crashes or even some low-speed crashes (Otte, Schroeder, and Richter, 2002; de Rome and Stanford, 2003).

Other types of protective gear include crash bars, motorcycle airbag jackets, and back and leg protectors. There is some evidence that crash bars protect motorcyclists' lower legs from side

---

[6] Directive 89/686/EEC was adopted on December 21, 1989, and became Community Law on July 1, 1992. To comply with the directive, motorcycle PPE must be independently tested and certified and be labeled with a Conformite Europeen label, which indicates that the PPE conforms to the relevant European standard (Motor Cycle Council of NSW, Inc., 2009).

impacts, but little evidence on the effectiveness of devices such as airbag jackets and back and leg protectors (Ross, 1983; Lin and Kraus, 2009).

NHTSA has recommended that motorcyclists should be educated about the benefits of PPE, such as eye protection, full coverage clothing, gloves, and boots (NHTSA and MSF, 2000, 2006). Conducting research on PPE effectiveness and developing or adopting existing standards for PPE was also recommended (NHTSA and MSF, 2000).

Safety Campaigns for Motorcycling

As discussed in Chapter Four, almost one-third of motorcyclist fatalities are caused in part by drinking. A 2009 literature review found that there were no existing programs that specifically attempted to reduce alcohol consumption by motorcycle riders, only general interventions that target all drivers (Lin and Kraus, 2009). The laws covering drinking and driving also apply to motorcyclists, but research has not found them to be particularly effective. A 2003 study looked at the association between alcohol-related automobile and motorcycle fatalities in the United States from 1980 to 1997 and several alcohol-related laws, including legal limits on BAC levels, zero tolerance (a BAC limit of 0.02 or less for persons younger than 21 years of age), administrative license revocation, sobriety checkpoints, and mandatory jail terms for first convictions for driving under the influence of alcohol. Only laws specifying a BAC legal limit of 0.08 were effective in reducing alcohol-related motorcycle fatalities. There was also a weak association between administrative license revocation laws and reductions in alcohol-related motorcycle fatalities (Villaveces et al., 2003). No other laws were effective in reducing alcohol-related motorcycle fatalities.

Some researchers have recommended a lower BAC limit for motorcycle riders. Colburn et al. (1993) recommended lower legal limits for BAC levels for motorcycle riders based on increased reaction time and performance errors in a simulator test by motorcyclists with BAC levels below the legal limit. Sun et al. (1998) looked at BAC levels of all drivers of motorcycle and motor vehicles admitted to a New Jersey Level I Trauma Center in 1992 and found the mean BAC level was lower in motorcycle riders involved in crashes than in car drivers. They recommended a BAC limit of 0.05 for motorcyclists because of the greater risk of injury from riding a motorcycle and the additional coordination and balance required for operating a two- versus a four-wheeled vehicle (Sun et al., 1998). Argentina has a lower BAC limit for motorcycle riders (0.02 percent) than for cars (0.05 percent) (quoted in Davis et al., 2003).

A 2001 paper looking specifically at fatal and severe injury motor vehicle crashes involving U.S. Air Force personnel also recommends lowering the overall BAC level for all military personnel, not just motorcyclists. Specifically, the paper states: "With a high concentration of at-risk youth in the military, a universal BAC limit of 0.05 percent, which follows the American Medical Association recommendation and is sensitive to motorcyclists, is a very sound concept, and therefore is recommended" (Carr, 2001, p. 33). Zero tolerance for military personnel aged 25 and younger is also suggested as a reasonable course of action (Carr, 2001).

CHAPTER EIGHT

Findings and Further Research Needs

As noted throughout this report, the demographic characteristics of service members mean that the military faces difficult challenges in the area of driver safety. While the percentage of fatal crashes among service members is comparable to or lower than that for men of similar ages outside of the military, it is substantially higher than what is observed in the general population. Driving risk is higher among young men (deaths in crashes occur at double the rate for the U.S. population overall), and as a consequence, a military force made up of such persons is at great risk for crash injury and death.

In addition, while the percentage of total crash deaths occurring on motorcycles is modest, the risk of crash death for riders versus drivers of passenger cars is far greater. Moreover, data show that motorcycle deaths are increasing (in contrast to the declining risk of passenger vehicle travel), and the number of fatal motorcycle crashes among service members has grown over the past decade. In the U.S. population, the percentage of motor vehicle deaths occurring on motorcycles has tripled, increasing from about 5 percent to about 15 percent in just over a decade. These percentages are higher in all of the armed services, and rates in these populations have also risen. In the Navy, Marines, and Coast Guard, the proportion of motor vehicle deaths occurring on motorcycles is presently close to 50 percent. As with passenger vehicle deaths, the high rates of motorcycle fatalities can likely be attributed, in part, to the high percentage of young men making up the service member population.

We identified a number of key predictors of crashes and crash death. Among military members, motorcycle riding, drinking, fatigue, and failure to wear a seat belt (or on motorcycles, a helmet) are all associated with an increased risk of accidental death or crash.[1] These risks are confirmed in studies of the general population, and these studies also point to a role for demographics in the risks observed for military personnel. In 2008, in 22 percent of fatal crashes in the United States drivers had BAC levels over the current legal limit. BAC levels were over the limit among 34 percent of young adult drivers in such crashes, and over-the-limit rates among males were nearly twice what they were among females. Among service members, rates of binge drinking (a level of consumption likely to put one over the BAC limit) are higher than in the general population, and rates of drinking and driving are also high. About three-quarters of all binge drinking episodes among active duty personnel appear to occur among those under the legal drinking age (21 years), although whether these same individuals also drink and drive is unknown.

[1] As discussed in Chapter Three, the studies of military personnel found conflicting evidence about the role of habitually speeding as a predictor of crashes.

Speeding and failure to use seat belts pose substantial risks as well. As many as 40 percent of crashes caused by male drivers 15 to 20 years old may be the result of speeding, and seat belt use is estimated to reduce the likelihood of death by 50 percent, given a vehicle crash. Young men are more apt to be involved in speed-related crashes than other demographic groups, and less likely to use seat belts. Fatigued and distracted driving are more difficult to assess because they are not tracked in police reports. They appear to contribute less to rates of vehicle crashes, but they are risky nonetheless, with estimates of involvement ranging between 10 and 20 percent. Cell phone use while driving is more common among young people.

Some additional factors come into play in motorcycle crash rates. Increases in the number of sport and supersport bikes on the road appear to play a key role in the increased rates of motorcycle crash death. These powerful, lightweight bikes are more often driven by younger drivers. In addition, conspicuity of cycles may be important in preventing crashes with vehicles of other types, although the research provides conflicting evidence at this juncture regarding the best ways to increase it.

In addition to younger age and male gender, factors that predict risky driving practices in both the general population and among military personnel include personality characteristics (particularly sensation-seeking), heavy or binge drinking, prior drinking and driving convictions, beliefs related to masculinity, risk perceptions, and perhaps deployment. In each of these cases, it is difficult to be certain of a causal role, but they provide important information for the design and targeting of preventive interventions to reduce traffic fatalities among military personnel.

The remainder of this chapter discusses these findings and how they might be translated into programs that could be adopted by the military. However, we first make two observations about reducing vehicle crash deaths. First, the United States has made a substantial amount of progress in reducing crash deaths over the past decades. A number of factors contributed to this: improvements in vehicle safety (such as seat belts), better emergency medicine, and changes in public opinion and behavior (such as an increased awareness of the dangers of drunk driving). All of these factors affect military crash deaths as well. These accomplishments have been maintained over time, as evidenced by the continued decrease in fatalities per VMT. However, the rate of progress seems to be slowing, and recent decreases in fatalities in 2008 and 2009 may be due largely to the decrease in miles driven as a result of the recession. It is possible that even with new safety interventions on the part of the military, it will be difficult to reduce crash deaths dramatically, simply because many improvements have already been realized.

Second, even given that context, there may be reason to think that some additional safety interventions may be effective, to the extent that the military can effect positive changes in its service members' behavior. The military may in some ways have an easier time of changing behavior, because it can mandate and punish behavior above and beyond existing state and federal laws; for example, the military requires helmet use when riding motorcycles, even though state laws vary. Of course, in both the civilian and military world, even when laws exist to prohibit certain behavior, it still occurs. Drunk driving still accounts for about one-third of all deaths in vehicle crashes, even though it is illegal. But to the extent that the military has a large degree of influence over behavior, there may be gains from some of the suggestions here.

We reviewed existing safety interventions and identified a number that are likely to be effective in addressing the needs of the military. If there were more studies specifically about military crashes, we might be able to make even more targeted suggestions, but as noted in the report there have been relatively few. Instead we have concentrated on the areas in which we can draw lessons from the civilian world, and on areas of particular importance to the military: issues of risk-taking, drunk driving, and motorcycle safety.

Programs and strategies that have worked well in the general population include media campaigns, particularly those that are carefully designed and based in theory, and then broadly and strategically disseminated. In particular, while media safety campaigns do exist in the military, one key finding of this report is that high sensation-seekers react differently than low sensation-seekers to the same messages, Because of the large proportion of high sensation-seekers in the military, we think it would be useful to design campaigns directed at this target audience. Changes in traffic regulations and enforcement have been shown to reduce injuries and death, with seat belt regulations and zero tolerance for drinking among underage drivers standing out in their effectiveness. High visibility enforcement campaigns have been widely used to address various traffic issues in the general population, and have also worked well. In addition, strategies targeting alcohol consumption (since drinking can easily lead to drinking and driving) have been effective at reducing crashes and fatalities. In general, research suggests that interventions to improve driver safety will be most effective if they are multipronged, involving two or more complementary approaches to addressing the problem. The effectiveness of each aspect of multipronged strategies appears to benefit from the presence of the others.

Based on these findings, we make several suggestions based on the findings in this report that may help reduce motor vehicle fatalities among military members. Each of these may be more effective if implemented in conjunction with others. All should be accompanied by a rigorous evaluation to determine whether the strategy is working and if not, what changes might be made to make it more effective. Several notes about these suggested policies:

- These are not listed in order of importance; it would be difficult to assess which might be most effective.
- We did not review existing military policies in these areas to develop this list, so many of the details of implementation would have to be determined by the military.
- Finally, we did not develop suggested policies for speeding, fatigued driving, or distracted driving, because the review did not locate any policies that had been rigorously evaluated and found effective.

Enforce zero tolerance of underage drinking and driving (that is, underage personnel caught driving with a BAC level of over 0.02 would be considered DWI) and widely publicize this new policy. Because binge drinking is strongly related to crash fatalities and to risky driving behavior, and because the vast majority of binge drinking episodes in the military occur among underage personnel, enforcement of the minimum legal drinking age for drivers is likely to have a strong impact on crashes and crash deaths. Although it is currently against policy for personnel under 21 years to drink, it has been suggested that this policy is rarely enforced, and that punishment is unlikely. Enhanced enforcement is likely to function largely through deterrence (not arrests), and thus it is essential that personnel be made aware that the probability of punishment has increased.

To deter drunk driving by those who live on military bases, adapt DWI checkpoint methods for use on bases. This is based on the data showing that such checkpoints, when accompanied by campaigns advertising their presence, have been highly effective in civilian environments. Checkpoints may reduce DWI among service members if they can be implemented in such a way that they are not easy to circumvent (i.e., not placed at the same or all entrances to a base at known times and dates) but are at the same time highly visible and well-publicized. Checkpoints work by deterring DWI, not by increasing arrests, thus, it is important that service members are aware of the checks, but find them unavoidable when returning to base. If placement of checkpoints under such specifications is difficult, an alternative is the use of roving or saturation patrols that detect possible DWI cases in an area (e.g., some section of a base where drinking and driving is common), rather than screening all drivers. These patrols have been found to be effective in Michigan, where sobriety checkpoints are not permitted (Fell et al., 2008).

Similarly, consider using the same high-visibility enforcement technique to enforce seat belt laws. While military personnel are required to wear seat belts, surveys cited in this report have not found universal compliance with this requirement. As noted in Chapter Three, military personnel who do not routinely wear seat belts are at greater risk of dying in a vehicle crash than those who do. High-visibility enforcement campaigns have successfully incorporated both drunk driving and seat belt use.

Consider adopting a lower permitted BAC limit for legal driving or operation of a motorcycle among military personnel than that permitted by states. For example, the military might consider automatically revoking or suspending the right of service members to drive at a BAC threshold of 0.05. Enforcement could be based on the driving record information already obtained by the military from states, or through the use of sobriety checkpoints on bases, as noted immediately above. If implemented, two factors will be critical to success. First, service members must feel certain that if they are caught driving over the BAC limit their licenses will be lost. Because there may be a perception among service members that drinking and driving laws are selectively enforced, punishment will need to be mandatory and generally known to be so. Second, careful communication will need to accompany any change in BAC limits, which is likely to be perceived by personnel as unfairly restrictive, given that the restrictions are stronger than for civilians. Linking the change to military values such as readiness and strength may be helpful in gaining support for a more restrictive policy.

Consider conducting screening and providing brief intervention for military personnel with alcohol problems that do not rise to the level of an alcohol dependence disorder. Screening could be accomplished as part of standard health care. Military personnel who have had a DWI could also be referred for such screening. Those who misuse alcohol could be referred for brief intervention, which can be delivered in either a healthcare or community setting, by physicians, counselors, or other individuals. Delivering a brief intervention requires the use of previously validated approaches and (typically) one to two days of training for the counselor involved. Those that have been used previously consist of a short counseling session and group discussions. RAND is currently developing and testing web-based brief intervention for individuals who have been convicted of DWI and court-referred for alcohol education. Experts in brief intervention approaches should be closely involved in deciding the approach best suited for the military, and in the initial development and implementation of a brief intervention program. Screening and brief intervention have the advantage that they may also reduce depression and PTSD symptoms, which often co-occur with alcohol misuse.

Conduct a well-designed media campaign in conjunction with base and community activities designed to highlight the importance of safe driving among military personnel. Such a campaign should be based in theory, link safe driving to core values among military members and be pre-tested for effectiveness among young men and sensation-seeking individuals. Such a campaign would increase awareness of safe driving, and shift norms among military members toward seeing safety as the expected standard and as typical behavior. The message should also be widely disseminated and placed in appropriate outlets to ensure it reaches the target audience of young sensation-seekers—especially as all studies of military drivers have found men in their late teens and early 20s at greatest crash risk—and does so repeatedly so that exposure is sufficient to produce behavior change. The media campaign should be complemented by activities and events that create an environment that supports change. For example, safe driving educational fairs and classes, alcohol-free social events publicized as such, or opportunities for service members to mentor younger personnel or community youth in driving safety. Combining these with some of the regulatory and enforcement changes suggested above would further strengthen environmental support for the changes in driving behavior promoted by the campaign, and make it more effective.

Consider requiring that motorcyclists own the vehicles they drive and increase enforcement (swift and certain punishment) of those who operate a motorcycle without a license. Increased enforcement might be accomplished through mandatory military punishments for violators. Violations could be detected via traffic record reports to the military by civilian police, and/or through random checks for licenses among drivers operating their vehicles on base. As with other efforts involving increased enforcement, it is important to implement the strategy as a deterrent to unsafe riding rather than a punishment for it, and thus any such changes in policy should be communicated broadly and repeatedly to personnel before and during implementation. This communication should also include messages likely to increase acceptance of the new policy, for example, by emphasizing increasing crash deaths among cyclists and the link between crashes and unlicensed motorcycle operation. Simultaneously offering training may also be useful. Although evidence regarding the effectiveness of motorcycle training is mixed, as noted elsewhere in this report, offering personnel a chance to become properly licensed at the same time that enforcement of licensing requirements is increased is likely to strengthen and reinforce the change in safety norms that each communicates.

It would also be useful for the military to conduct further research to examine the factors that contribute specifically to the risk of traffic crashes or death among military personnel. While we are not fully able to scope such studies in this report, six studies might be useful in this regard:

- Research on the specific locations and causes of crashes. Mapping the crashes that occur at or near a particular base might shed light on whether certain roads or intersections are particularly risky (which might in turn lead a base to work with local authorities on safety improvements), or in the case of drunk driving crashes, which bars are frequented by service members (which might lead to bartender training at particular establishments). Analyzing the causes of crashes might identify specific issues to be addressed (for example, Radun et al. [2007] found that many crashes occurred when conscripts were driving home fatigued after a week of training in the forest). Such research might determine whether fatigue or suicide are causes of some crashes, which would suggest a different set of safety interventions. Or it might find that many service members were killed in crashes through no fault of their own.
- Research on the effects of deployment, stress, substance abuse, and sensation-seeking as predictors of risky driving. Although we found a number of military-specific studies in our review, few were comprehensive enough to allow us to determine which among a

broad set of factors poses the greatest risk and thus is the most likely target for intervention. Nor was it clear how various factors might interact to produce risk. For example, further research may allow us to tease apart the contributions of deployment, suicide, stress, and drinking to risky driving and crashes, or the roles of risk perception, sensation-seeking, and stereotypes of masculinity as involving risky driving. In particular, research might be designed to look at patterns of drinking and driving by age group, and how risk factors differ for car/truck drivers versus motorcyclists. It may be useful to add a broader set of items to existing surveys of health behavior among military personnel, or conduct supplemental surveys of these individuals. Survey responses could be matched with records of crashes to determine those service members most at risk.

- Research on when motorcycle training is effective in reducing crashes and when it might promote greater risk. As Chapter Seven notes, the civilian studies that have been conducted about whether motorcycle training is effective have found very mixed results, ranging from effective to no effect to increasing the risk of crashing. This is an area where the military could contribute to the understanding of the type of training and the circumstances under which it is effective. Perhaps new riders who have taken the first required round of training could receive supplemental training of various types or at differing intervals, and their skill at handling a motorcycle compared by trained observers and their propensity for risk-taking measured by surveys. Research might also look at the effects of training in conjunction with the type of bikes, including sport and supersport bikes.
- Research on the effectiveness of the military's existing safety programs. As evidenced by the wide variety of initiatives undertaken by the PMV TF, the military is already trying a number of programs to reduce preventable crashes. However, the evidence for their success or lack thereof seems to be largely anecdotal, and determined largely by the total annual numbers of vehicle crash deaths without controlling for other factors. The task force might seek to more rigorously examine some of its initiatives, such as Track Day for motorcycle training or the effectiveness of particular public service advertisements.
- Research on determining the motorcycle rider population across the services. It is important to identify this population for two reasons: first, motorcyclists are at far greater risk of being killed in a vehicle crash than those in cars. Second, as evidenced in at least one study, it seems that the population of those who are at highest risk for motorcycle crashes is demographically distinct from those at risk in car crashes. Having this information available across the services in a consistent format may assist in developing targeted safety measures, including appropriate media campaigns.
- Research on the prevalence of fatigue and distracted driving in military crashes. There is a wide range of estimates in the civilian population, and it is possible that the military profile may be different. Research on non-U.S. military populations has found fatigue to be a significant contributing factor to fatal crashes (Radun and Radun, 2009). This information may assist with developing an appropriate media campaign or in reviewing policies that may contribute to fatigued or distracted driving.

APPENDIX

Military Crash Deaths

Table A.1 shows the raw data we obtained for motor vehicle crash deaths for each service. Not all data were available for each service.

Table A.1
Military Fatalities from Motor Vehicle Crashes, 2000–2009

	2000	2001	2002	2003	2004	2005	2006	2007	2008	2009	Average
Army											
Car fatalities	92	79	84	90	110	104	79	77	78	78	
Motorcycle fatalities	12	15	26	19	22	40	49	38	51	32	
Total PMV fatalities	104	94	110	109	132	144	128	115	129	110	
Rate per 100,000 Army population	18.0	16.0	18.0	16.0	19.0	21.0	19.0	17.0	18.0	15.0	17.7
Air Force											
Car fatalities	26	36	52	44	38	30	25	28	14	27	
Motorcycle fatalities	8	9	19	23	22	15	17	18	15	20	
Total PMV fatalities	34	45	71	67	60	45	42	46	29	47	
Rate per 100,000 Air Force population	8.3	10.9	17.2	16.2	14.5	11.0	10.2	11.5	7.5	12.1	11.9
Navy											
Car fatalities	39	44	54	41	42	33	50	33	30	18	
Motorcycle fatalities	13	11	15	23	25	22	27	19	33	13	
Total PMV fatalities (includes pedestrian/bicycle)	53	57	76	67	73	60	79	53	67	34	
Rate per 100,000 Navy population	13.5	15.0	18.4	16.4	17.9	15.3	20.3	14.1	18.4	9.8	15.9
% car fatalities due to alcohol	NA	NA	39%	49%	33%	52%	38%	39%	40%	22%	39%
% motorcycle fatalities due to alcohol	NA	NA	27%	22%	32%	27%	11%	26%	30%	8%	23%
Marines											
Car fatalities	51	25	50	34	35	29	41	34	23	26	
Motorcycle fatalities	7	6	11	17	7	13	16	19	25	14	
Total PMV fatalities (includes pedestrian/bicycle)	61	33	65	53	46	45	64	57	51	43	

Table A.1—Continued

	2000	2001	2002	2003	2004	2005	2006	2007	2008	2009	Average
Rate per 100,000 Marine population	34.0	19.0	35.3	26.8	23.5	23.4	33.0	29.5	26.2	20.3	27.1
% car fatalities due to alcohol	NA	NA	20%	18%	37%	14%	32%	50%	39%	42%	31%
% motorcycle fatalities due to alcohol	NA	NA	9%	6%	14%	8%	6%	21%	16%	14%	12%
Coast Guard											
Car fatalities				6	3	3	5	3	5	1	
Motorcycle fatalities				4	1	2	3	4	4	2	
Total PMV fatalities				10	4	5	8	7	9	3	
Rate per 100,000 Coast Guard population	24.2	25.0	18.0	22.3	9.5	14.1	21.1	21.2	21.2	NA	19.6
Total U.S. population											
Car fatalities	33,451	33,243	34,105	33,627	33,276	33,070	32,119	30,527	26,689	NA	
Motorcycle fatalities	2,897	3,197	3,270	3,714	4,028	4,576	4,837	5,154	5,290	NA	
Total PMV fatalities (includes pedestrian/bicycle)	41,945	42,196	43,005	42,884	42,836	43,510	42,708	41,259	37,261	33,963[a]	
Fatalities per 100,000 population	14.9	14.8	15.0	14.8	14.6	14.7	14.3	13.7	12.3	11.1[a]	14.7
% car drivers in fatal crashes with BAC over 0.08			23%	23%	22%	22%	23%	23%	23%	NA	23%
% motorcyclists in fatal crashes with BAC over 0.08			31%	29%	27%	27%	27%	27%	29%	NA	28%

SOURCES: Army data: Total fatalities and fatality rate, ASMIS, 2010; motorcycle fatalities, Walter Beckman of U.S. Army Combat Readiness/Safety Center, email to Liisa Ecola on February 16, 2010. Air Force data: Terry L. Todd, Air Force Senior Master Sergeant, email to Liisa Ecola on January 4, 2010. Navy/Marine data: Naval Safety Center, 2010. Coast Guard data: Dale Wisnieski of the U.S. Coast Guard, email to Liisa Ecola on February 17, 2010. U.S. population data: Fatalities from NHTSA, 2009c, except 2009 estimate from NCSA, 2010; proportion of deaths caused by drivers with over 0.08 BAC level, NHTSA, 2002, 2003, 2004, 2005a, 2006a, 2008b, and 2009a.

NOTE: NA is "not available."

[a] Total is estimated.

References

Aarts, Letty, and Ingrid van Schagen, "Driving Speed and the Risk of Road Crashes: A Review," *Accident Analysis and Prevention*, Vol. 38, No. 2, March 2006, pp. 215–224.

Abdel-Aty, M. A., and H. T. Abdelwahab, "Exploring the Relationship between Alcohol and the Driver Characteristics in Motor Vehicle Accidents," *Accident Analysis and Prevention*, Vol. 32, No. 4, 2000, pp. 473–482.

American Motorcyclist Association, *Motorcycle Laws by State*, Pickerington, Ohio: American Motorcyclist Association, March 29, 2010.

Ames, Genevieve, and Carol Cunradi, "Alcohol Use and Preventing Alcohol-Related Problems Among Young Adults in the Military," *Alcohol Research and Health*, Vol. 28, No. 4, 2004/2005, pp. 252–257.

Araujo, Marcus Maximilliano, Leandro Fernandes Malloy-Diniz, and Fábio Lopes Rocha, "Impulsividade e acidentes de trânsito," [Impulsiveness and Traffic Accidents], *Revista De Psiquiatria Clinica*, Vol. 36, No. 2, 2009, pp. 54–62.

Arnett, Jeffrey Jenson, Daniel Offer, and Mark A. Fine, "Reckless Driving in Adolescence: 'State' and 'Trait' Factors," *Accident Analysis and Prevention*, Vol. 29, No. 1, 1997, pp. 57–63.

ASMIS—*see* U.S. Army Safety Management Information System

Baer, Justin, and Melanie Skerner, *Review of State Motorcycle Safety Program Assessments*, Washington, D.C.: National Highway Traffic Safety Administration, U.S. Department of Transportation, DOT HS 811 082, January 2009.

Baer, Justin D., Andrea L. Cook, and Stephane Baldi, *Motorcycle Rider Education and Licensing: A Review of Programs and Practices*, Washington, D.C.: National Highway Traffic Safety Administration, U.S. Department of Transportation, DOT HS 809 852, 2005. As of January 28, 2010: http://www.nhtsa.dot.gov/people/injury/pedbimot/motorcycle/McycleRiderWeb/pages/index.htm

Baldi, S., J. D. Baer, and A. L. Cook, "Identifying Best Practices States in Motorcycle Rider Education and Licensing," *Journal of Safety Research*, Vol. 36, No. 1, 2005, pp. 19–32.

Bell, Nicole S., Paul J. Amoroso, David H. Wegman, and Laura Senier, "Proposed Explanations for Excess Injury Among Veterans of the Persian Gulf War and a Call for Greater Attention from Policymakers and Researchers," *Injury Prevention*, Vol. 7, No. 1, March 2001, pp. 4–9.

Bell, Nicole S., Paul J. Amoroso, Jeffrey O. Williams, Michelle M. Yore, Charles C. Engel Jr., Laura Senier, Annette C. DeMattos, and David H. Wegman, "Demographic, Physical, and Mental Health Factors Associated with Deployment of U.S. Army Soldiers to the Persian Gulf," *Military Medicine*, Vol. 165, No. 10, 2000a, p. 762.

Bell, Nicole S., Paul J. Amoroso, Michelle M. Yore, Gordon S. Smith, and Bruce H. Jones, "Self-Reported Risk-Taking Behaviors and Hospitalization for Motor Vehicle Injury Among Active Duty Army Personnel," *American Journal of Preventive Medicine*, Vol. 18, No. 3, April 2000b, pp. 85–95.

Billheimer, John W., "Evaluation of California Motorcyclist Safety Program," *Transportation Research Record: Journal of the Transportation Research Board*, Vol. 1640, 1998, pp. 100–109.

Bingham, C. Raymond, and Jean T. Shope, "Adolescent Developmental Antecedents of Risky Driving Among Young Adults," *Journal of Studies on Alcohol*, Vol. 65, No. 1, 2004a.

———, "Adolescent Problem Behavior and Problem Driving in Young Adulthood," *Journal of Adolescent Research*, Vol. 19, No. 2, 2004b, pp. 205–223.

Blanchard, Howard Thomas, and Patricia A. Tabloski, "Motorcycle Safety: Educating Riders at the Teachable Moment," *Journal of Emergency Nursing*, Vol. 32, No. 4, August 2006, pp. 330–332.

Blows, Stephanie, Shanthi Ameratunga, Rebecca Q. Ivers, Sing Kai Lo, and Robyn Norton, "Risky Driving Habits and Motor Vehicle Driver Injury," *Accident Analysis and Prevention*, Vol. 37, No. 4, 2005, pp. 619–624.

Boehmer, T. K., W. D. Flanders, M. A. McGeehin, C. Boyle, and D. H. Barrett, "Postservice Mortality in Vietnam Veterans: 30-Year Follow-Up," *Archives of Internal Medicine*, Vol. 164, No. 17, September 27, 2004, pp. 1908–1916.

Boning, W. Brent, and Michael D. Bowes, *Statistical Analysis of USMC Accidental Deaths*, Alexandria, Va.: Center for Naval Analyses, CRM D0008914.A1/SR1, 2003.

Bowes, Michael D., and Catherine M. Hiatt, *An Analysis of USMC Accidental Deaths: 2007 Update*, Alexandria, Va.: Center for Naval Analyses, CRM D0018760.A2/Final, 2008.

Bowie, N. N., Jr., and M. Waltz, "Data Analysis of the Speed-Related Crash Issue," *Auto and Traffic Safety*, Vol. 2, Winter 1994.

Boyle, C. A., P. Decoufle, R. J. Delaney, F. DeStefano, M. L. Flock, M. I. Hunter, R. Joesoef, J. M. Karon, M. L. Kirk, and P. M. Layde et al., "Postservice Mortality Among Vietnam Veterans," *Journal of the American Medical Association*, Vol. 257, No. 6, February 1987, pp. 790–795.

Bray, Robert M., J. A. Fairbank, and Mary Ellen Marsden, "Stress and Substance Use Among Military Women and Men," *American Journal of Drug & Alcohol Abuse*, Vol. 25, 1999, pp. 239–256.

Bray, Robert M., and Laurel L. Hourani, "Substance Use Trends Among Active Duty Military Personnel: Findings from the United States Department of Defense Health Related Behavior Surveys, 1980–2005," *Addiction*, Vol. 102, 2007, pp. 1092–1101.

Bray, Robert M., Michael R. Pemberton, Laurel L. Hourani, Michael Witt, Kristine L. Rae Olmsted, Janice M. Brown, BeLinda Weimer, Marian E. Lane, Mary Ellen Marsden, Scott Scheffler, Russ Vandermaas-Peeler, Kimberly R. Aspinwall, Erin Anderson, Karthryn Spagnola, Kelly Close, Jennifer L. Gratton, Sara Calvin, and Michael Bradshaw, *2008 Department of Defense Survey of Health Related Behaviors Among Active Duty Military Personnel: A Component of the Defense Lifestyle Assessment Program (DLAP)*, Research Triangle Park, N.C.: RTI International, RTI/10940-FR, September 2009.

Brewer, Robert D., Peter D. Morris, Thomas B. Cole, Stephanie Watkins, Michael J. Patetta, and Carol Popkin, "The Risk of Dying in Alcohol-Related Automobile Crashes Among Habitual Drunk Drivers," *New England Journal of Medicine*, Vol. 331, August 25, 1994, pp. 513–517.

Bureau of Transportation Statistics (BTS), *National Transportation Statistics*, Washington, D.C.: Research and Innovative Technology Administration, U.S. Department of Transportation, 2009. As of November 20, 2009: http://www.bts.gov/publications/national_transportation_statistics/

Burns, Peter C., and Gerald J. S. Wilde, "Risk Taking in Male Taxi Drivers: Relationships Among Personality, Observational Data and Driver Records," *Personality and Individual Differences*, Vol. 18, No. 2, 1995, pp. 267–278.

Caetano, R., and C. L. Clark, "Hispanics, Blacks and Whites Driving Under the Influence of Alcohol: Results from the 1995 National Alcohol Survey," *Accident Analysis and Prevention*, Vol. 32, No. 1, 2000, pp. 57–64.

Caird, Jeffrey K., Charles Scialfa, T., Geoffrey Ho, and Alison Smiley, *Effects of Cellular Telephones on Driving Behaviour and Crash Risk: Results of Meta-Analysis*, Quebec, Canada: CAA Foundation for Traffic Safety, 2004. As of June 25, 2010:
http://www.ama.ab.ca/images/images_pdf/FinalReport_CellPhones4.pdf

Carr, Bridget K., *Fatal and Severe Injury Motor Vehicle Crashes Involving Air Force Personnel*, Montgomery, Ala.: Maxwell Air Force Base, Air Command and Staff College, Air University, 2001. As of January 1, 2010: http://www.dtic.mil/cgi-bin/GetTRDoc?AD=ADA407057&Location=U2&doc=GetTRDoc.pdf/

Castella, J., and J. Perez, "Sensitivity to Punishment and Sensitivity to Reward and Traffic Violations," *Accident Analysis and Prevention*, Vol. 36, No. 6, November 2004, pp. 947–952.

Centers for Disease Control and Prevention (CDC), *Wonder on-Line Database*, "Compressed Mortality File, 1999–2006," Series 20, No. 2L, Hyattsville, Md.: CDC, 2009. As of June 25, 2010:
http://wonder.cdc.gov/cmf-icd10.html

———, "National Vital Statistics System—U.S. Census Populations with Bridged Race Estimates," Hyattsville, Md.: CDC, 2003. As of June 1, 2010:
http://www.cdc.gov/nchs/nvss/bridged_race.htm

Chappell, Bill, "Scooter Rundown: Best Fits from Tall to Small," Washington, D.C.: NPR, August 21, 2008. As of January 25, 2009:
http://www.npr.org/templates/story/story.php?storyId=93434949

Cherpitel, Cheryl J., "Substance Use, Injury, and Risk-Taking Dispositions in the General Population," *Alcoholism: Clinical and Experimental Research*, Vol. 23, No. 1, 1999, pp. 121–126.

Christie, S. M., R. A. Lyons, F. D. Dunstan, and S. J. Jones, "Are Mobile Speed Cameras Effective? A Controlled Before and After Study," *Injury Prevention*, Vol. 9, No. 4, December 2003, pp. 302–306.

Cismaru, M., A. M. Lavack, and E. Markewich, "Social Marketing Campaigns Aimed at Preventing Drunk Driving: A Review and Recommendations," *International Marketing Review*, Vol. 26, No. 3, 2009, pp. 292–311.

Colburn, N., R.D. Meyer, M. Wrigley, and E. L. Bradley, "Should Motorcycles Be Operated Within the Legal Alcohol Limits for Automobiles," *Journal of Trauma*, Vol. 35, No. 2, 1993, pp. 183–186.

Creaser, Janet, Nic Ward, Mick Rakauskas, E. Boer, Craig Shankwitz, and Flavia Nardi, *Effects of Alcohol on Motorcycle Riding Skills*, Minneapolis, Minn.: University of Minnesota, Center for Transportation Studies, 2007. As of December 22, 2009:
http://www.cts.umn.edu/Publications/ResearchReports/reportdetail.html?id=1624/

CTIA the Wireless Association, "Wireless Quick Facts: Mid-Year Figures," Washington, D.C.: CTIA the Wireless Association, 2010. As of January 18, 2010:
http://www.ctia.org/media/industry_info/index.cfm/AID/10323/

D'Amico, Elizabeth, and Maria Orlando Edelen, "Pilot Test of Project Choice: A Voluntary After School Intervention for Middle School Youth," *Psychology of Addictive Behaviors*, Vol. 21, No. 4, 2007, pp. 592–598.

Davis, A., A. Quimby, W. Odero, G. Gururaj, and M. Hijar, *Improving Road Safety by Reducing Impaired Driving in Developing Countries: A Scoping Study*, Crowthorne, Berkshire, UK: Global Road Safety Partnership and Department for International Development, PR/INT/724/03, 2003.

de Hoog, Natascha, Wolfgang Stroebe, and John B. F. de Wit, "The Impact of Vulnerability to and Severity of a Health Risk on Processing and Acceptance of Fear-Arousing Communications: A Meta-Analysis," *Review of General Psychology*, Vol. 11, No. 3, September 2007, pp. 258–285.

de Rome, Liz, and Guy Stanford, *Motorcycle Protective Clothing*, Colyton, New South Wales, Australia: Motor Accidents Authority of NSW, 2003.

de Wit, John B. F., Enny Das, and Raymond Vet, "What Works Best: Objective Statistics or a Personal Testimonial? An Assessment of the Persuasive Effects of Different Types of Message Evidence on Risk Perception," *Health Psychology*, Vol. 27, No. 1, January 2008, pp. 110–115.

Dill, Patricia L., Elisabeth Wells-Parker, and Carl A. Soderstrom, "The Emergency Care Setting for Screening and Intervention for Alcohol Use Problems Among Injured and High-Risk Drivers: A Review," *Traffic Injury Prevention*, Vol. 5, No. 3, 2004, pp. 278–291.

Dingus, T. A., S. G. Klauer, V. L. Neale, A. Petersen, S. E. Lee, J. Sudweeks, M. A. Perez, J. Hankey, D. Ramsey, S. Gupta, C. Bucher, Z. R. Doerzaph, J. Jermeland, and R. R. Knipling, *The 100-Car Naturalistic Driving Study: Phase II—Results of the 100-Car Field Experiment*, Washington, D.C.: National Highway Traffic Safety Administration, U.S. Department of Transportation, DOT HS 810 593, April 2006.

Drews, Frank A., Monisha Pasupathi, and David L. Strayer, "Passenger and Cell Phone Conversations in Simulated Driving," *Journal of Experimental Psychology—Applied*, Vol. 14, No. 4, 2008, pp. 392–400.

Drews, Frank A., Hina Yazdani, Celeste N. Godfrey, and Joel M. Cooper, "Text Messaging During Simulated Driving," *Human Factors: The Journal of the Human Factors and Ergonomics Society*, Vol. 51, No. 5, 2009.

Dussault, C., "Guidelines for Enforcement Campaigns," *Health Education Research*, Vol. 5, No. 2, 1990, pp. 217–223.

Elander, James, Robert West, and Davina French, "Behavioral Correlates of Individual Differences in Road-Traffic Crash Risk: An Examination of Methods and Findings," *Psychological Bulletin*, Vol. 113, No. 2, 1993, pp. 279–294.

Elder, Randy W., Ruth A. Shults, David A. Sleet, James L. Nichols, Robert S. Thompson, and Warda Rajab, "Effectiveness of Mass Media Campaigns for Reducing Drinking and Driving and Alcohol-Involved Crashes—a Systematic Review," *American Journal of Preventive Medicine*, Vol. 27, No. 1, July 2004, pp. 57–65.

Elvik, R., "Can Injury Prevention Efforts Go Too Far? Reflections on Some Possible Implications of Vision Zero for Road Accident Fatalities," *Accident Analysis and Prevention*, Vol. 31, No. 3, 1999, pp. 265–286.

Fear, Nicola T., Amy C. Iversen, Amit Chatterjee, Margaret Jones, Neil Greenberg, Lisa Hull, Roberto J. Rona, Matthew Hotopf, and Simon Wessely, "Risky Driving Among Regular Armed Forces Personnel from the United Kingdom," *American Journal of Preventive Medicine*, Vol. 35, No. 3, September 2008, pp. 230–236.

Federal Highway Administration, *National Household Travel Survey*, Washington, D.C.: Federal Highway Administration, 2001. As of June 25, 2010:
http://nhts.ornl.gov/

Federation of European Motorcyclists, *Rider Training in Europe—The Views and the Needs of the Rider, a Report on Initial Rider Training with Recommended Guiding Principles*, September 1997.

Fell, James C., Elizabeth A. Langston, John H. Lacey, A. Scott Tippetts, and Ray Cotton, Evaluation of Seven Publicized Enforcement Demonstration Programs to Reduce Impaired Driving: Georgia, Louisiana, Pennsylvania, Tennessee, Texas, Indiana, and Michigan: United States. National Highway Traffic Safety Administration, 2008. As of July 26, 2010:
http://ntl.bts.gov/lib/30000/30200/30205/810941.pdf

Fergusson, David, Nicola Swain-Campbell, and John Horwood, "Risky Driving Behaviour in Young People: Prevalence, Personal Characteristics and Traffic Accidents," *Australian and New Zealand Journal of Public Health*, Vol. 27, No. 3, 2003, pp. 337–342.

Ferrier-Auerbach, Amanda G., Shannon M. Kehle, Christopher R. Erbes, Paul A. Arbisi, Paul Thuras, and Melissa A. Polusny, "Predictors of Alcohol Use Prior to Deployment in National Guard Soldiers," *Addictive Behaviors*, Vol. 34, No. 8, August 2009, pp. 625–631.

Fishbein, M., and I. Ajzen, *Belief, Attitude, Intention and Behaviour: An Introduction to Theory and Research*, Reading, Mass.: Addison-Wesley, 1975.

French, Michael T., Gulcin Gumus, and Jenny F. Homer, "Public Policies and Motorcycle Safety," *Journal of Health Economics*, Vol. 28, No. 4, 2009, pp. 831–838.

Gackstetter, G. D., T. I. Hooper, S. F. DeBakey, A. Johnson, B. E. Nagaraj, J. M. Heller, and H. K. Kang, "Fatal Motor Vehicle Crashes Among Veterans of the 1991 Gulf War and Exposure to Munitions Demolitions at Khamisiyah: A Nested Case-Control Study," *American Journal of Industrial Medicine*, Vol. 49, No. 4, April 2006, pp. 261–270.

Gerard, David, Paul S. Fischbeck, Barbara Gengler, and Randy S. Weinberg, "An Interactive Tool to Compare and Communicate Traffic Safety Risks: Trafficstats," paper presented at Annual Meeting of the Transportation Research Board, Washington, D.C., 2007.

Grube, J. W., "Alcohol Regulation and Traffic Safety: An Overview," in *Traffic Safety and Alcohol Regulations: A Symposium*, Transportation Research Circular No. E-C123, 2007.

Gullone, E., S. Moore, S. Moss, and C. Boyd, "The Adolescent Risk-Taking Questionnaire," *Journal of Adolescent Research*, Vol. 15, 2000, pp. 231–250.

Hagenzieker, Marjan P., "Enforcement or Incentives? Promoting Safety Belt Use Among Military Personnel in the Netherlands," *Journal of Applied Behavior Analysis*, Vol. 24, No. 1, Spring 1991, pp. 23–30.

Hanchulak, Denise, and Brett Robinson, *Guidelines for Motorcycle Operator Licensing*, Washington, D.C.: National Highway Traffic Safety Administration, DOT HS 811 141, 2009.

Hartos, Jessica, Patricia Eitel, and Bruce Simons-Morton, "Parenting Practices and Adolescent Risky Driving: A Three-Month Prospective Study," *Health Education and Behavior*, Vol. 29, No. 2, April 2002, pp. 194–206.

Haworth, Narelle, Rob Smith, and Naomi Kowadlo, *Evaluation of Rider Training Curriculum in Victoria*, Victoria, Australia: Monash University Accident Research Centre, Report No. 165, 2000.

Haworth, Narelle, and Christine Mulvihill, *Review of Motorcycle Licensing and Training*, Victoria, Australia: Monash University Accident Research Centre, Report No. 240, June 2005.

Hedlund, James, *Motorcyclist Traffic Fatalities by State: 2009 Preliminary Data*, Washington, D.C.: Governors Highway Safety Association, 2010.

Hedlund, J. H., R. G. Ulmer, and D. F. Preusser, *Determine Why There Are Fewer Young Alcohol-Impaired Drivers*, Washington, D.C.: National Highway Traffic Safety Administration, 2001.

Heino, Adriaan, Hugo H. van der Molen, and Gerald J. S. Wilde, "Differences in Risk Experience Between Sensation Avoiders and Sensation Seekers," *Personality and Individual Differences*, Vol. 20, No. 1, 1996, pp. 71–79.

Henderson, Robert L., Kenneth Ziedman, William J. Burger, and Kevin E. Cavey, *Motor Vehicle Conspicuity*, Warrendale, Pa.: Society of Automotive Engineers, SAE Technical Paper Series 830566, 1983.

Heron, Melonie, Donna L. Hoyert, Sherry L. Murphy, Jiaquan Xu, Kenneth D. Kochanek, and Betzaida Tejada-Vera, *Deaths: Final Data for 2006*, Centers for Disease Control and Prevention, National Vital Statistics Reports, Vol. 57, No. 14, April 17, 2009.

Highway Loss Data Institute, "Hand-Held Cellphone Laws and Collision Claim Frequencies," *Highway Loss Data Institute Bulletin*, Vol. 26, No. 17, December 2009.

Hingson, R., T. Heeren, and M. Winter, "Lower Legal Blood Alcohol Limits for Young Drivers," *Public Health Report*, Vol. 109, No. 6, 1994, pp. 738–744.

Hooper, Tomoko I., S. F. DeBakey, K. S. Bellis, H. K. Kang, D. N. Cowan, A. E. Lincoln, and G. D. Gackstetter, "Understanding the Effect of Deployment on the Risk of Fatal Motor Vehicle Crashes: A Nested Case-Control Study of Fatalities in Gulf War Era Veterans, 1991–1995," *Accident Analysis and Prevention*, Vol. 38, No. 3, May 2006, pp. 518–525.

Hooper, Tomoko I., S. F. DeBakey, A. Lincoln, H. K. Kang, D. N. Cowan, and G. D. Gackstetter, "Leveraging Existing Databases to Study Vehicle Crashes in a Combat Occupational Cohort: Epidemiologic Methods," *American Journal of Industrial Medicine*, Vol. 48, No. 2, August 2005, pp. 118–127.

Hornik, R. C., ed. *Public Health Communication: Evidence for Behavior Change*, Mahwah, N.J.: Lawrence Erlbaum, 2002.

Horrey, William, and Christopher Wickens, "Examining the Impact of Cell Phone Conversations on Driving Using Meta-Analytic Techniques," *Human Factors*, Vol. 38, No. 1, 2006, pp. 196–205.

Hosking, Simon, Kristie Young, and Michael Regan, *The Effects of Text Messaging on Young Novice Driver Performance*, Victoria, Australia: Monash University Accident Research Center, 2006.

Houston, David J., and Lilliard E. Richardson, "Motorcycle Safety and the Repeal of Universal Helmet Laws," *American Journal of Public Health*, Vol. 97, No. 11, 2007, pp. 2063–2069.

Hurt, H. H., Jr., J. V. Ouellet, and D. R. Thom, *Motorcycle Accident Cause Factors and Identification of Countermeasures, Volume 1: Technical Report*, Washington, D.C.: National Highway Traffic Safety Administration, U.S. Department of Transportation, January 1981. As of December 20, 2009: http://isddc.dot.gov/OLPFiles/NHTSA/013695.pdf

Ingre, M., T. Akerstedt, B. Peters, A. Anund, G. Kecklund, and A. Pickles, "Subjective Sleepiness and Accident Risk, Avoiding the Ecological Fallacy," *Journal of Sleep Research*, Vol. 15, No. 2, June 2006, pp. 142–148.

Insurance Institute for Highway Safety (IIHS), "Special Issue: Motorcycles," *Status Report*, Vol. 42, No. 9, September 11, 2007.

———, "Current U.S. Motorcycle and Bicycle Helmet Laws," Arlington, Va.: IIHS, 2010a. As of July 16, 2010:
http://www.iihs.org/laws/HelmetUseCurrent.aspx

———, "Licensing Ages and Graduated Licensing Systems," Arlington, Va.: IIHS, 2010b. As of July 16, 2010:
http://www.iihs.org/laws/graduatedLicenseIntro.aspx

Irwin, C. E., and B. L. Halpern-Felsher, "Discussion Paper," *Injury Prevention*, Vol. 8, 2002, pp. ii21–ii23.

Ivers, Rebecca, Teresa Senserrick, Soufiane Boufous, Mark Stevenson, Huei-Yang Chen, Mark Woodward, and Robyn Norton, "Novice Drivers' Risky Driving Behavior, Risk Perception, and Crash Risk: Findings from the Drive Study," *American Journal of Public Health*, Vol. 99, No. 9, 2009, pp. 1638–1644.

Jacobson, Isabel G., Margaret A. K. Ryan, Tomoko I. Hooper, Tyler C. Smith, Paul J. Amoroso, Edward J. Boyko, Gary D. Gackstetter, Timothy S. Wells, and Nicole S. Bell, "Alcohol Use and Alcohol-Related Problems Before and After Military Combat Deployment," *Journal of the American Medical Association*, Vol. 300, No. 6, August 2008, pp. 663–675.

Jessor, R., and S. L. Jessor, *Problem Behavior and Psychosocial Development: A Longitudinal Study*, New York: Academic Press, 1977.

Johnson, Mark B., "The Consequences of Providing Drinkers with BAC Information on Assessments of Alcohol Impairment and Drunk-Driving Risk," paper presented at Young Impaired Drivers: The Nature of the Problem and Possible Solutions, A Workshop, Woods Hole, Mass., June 3–4, 2009.

Joksch, H. C., "Velocity Change and Fatality Risk in a Crash—A Rule of Thumb," *Accident Analysis and Prevention*, Vol. 25, No. 1, 1993.

Jonah, B. A., "Sensation Seeking and Risky Driving: A Review and Synthesis of the Literature," *Accident Analysis and Prevention*, Vol. 29, No. 5, September 1997, pp. 651–665.

Jonah, B. A., R. Thiessen, and E. Au-Yeung, "Sensation Seeking, Risky Driving, and Behavioral Adaptation," *Accident Analysis and Prevention*, Vol. 33, 2001, pp. 679–684.

Kahane, C. J., *Fatality Reduction by Safety Belts for Front-Seat Occupants of Cars and Light Trucks*, Washington, D.C.: U.S. Department of Transportation, DOT HS 809 199, 2000.

Kalsher, Michael J., E. Scott Geller, Steven W. Clarke, and Galen R. Lehman, "Safety Belt Promotion on a Naval Base: A Comparison of Incentives Vs. Disincentives," *Journal of Safety Research*, Vol. 20, No. 3, Fall, 1989, pp. 103–113.

Kang, Han K., and T. A. Bullman, "Mortality Among US Veterans of the Persian Gulf War: 7-Year Follow-Up," *American Journal of Epidemiology*, Vol. 154, No. 5, September 2001, pp. 399–405.

Kang, Han K., C.A. Magee, G. J. Macfarlane, and G. C. Gray, "Mortality Among U.S. and UK Veterans of the Persian Gulf War: A Review," *Occupational and Environmental Medicine*, Vol. 59, 2002, pp. 794–799.

Kessler, Ronald C., Patricia Berglund, Olga Demler, Robert Jin, Doreen Koretz, Kathleen R. Merikangas, A. John Rush, Ellen E. Walters, and Philip S. Wang, "The Epidemiology of Major Depressive Disorder: Results from the National Comorbidity Survey Replication (NCS-R)," *Journal of the American Medical Association*, Vol. 289, No. 23, June 18, 2003, pp. 3095–3105.

Kessler, Ronald C., Wai Tat Chiu, Olga Demler, and Ellen E. Walters, "Prevalence, Severity, and Comorbidity of 12-Month DSM-IV Disorders in the National Comorbidity Survey Replication," *Archives of General Psychiatry*, Vol. 62, No. 6, June 1, 2005, pp. 617–627.

Killgore, William D. S., Dave I. Cotting, Jeffrey L. Thomas, Anthony L. Cox, Dennis McGurk, Alexander H. Vo, Carl A. Castro, and Charles W. Hoge, "Post-Combat Invincibility: Violent Combat Experiences Are Associated with Increased Risk-Taking Propensity Following Deployment," *Journal of Psychiatric Research*, Vol. 42, No. 13, October 2008, pp. 1112–1121.

Klauer, S. G., T. A. Dingus, V. L. Neale, J. D. Sudweeks, and D. J. Ramsey, *The Impact of Driver Inattention on near-Crash/Crash Risk: An Analysis Using the 100-Car Naturalistic Driving Study Data*, Washington, D.C.: National Highway Traffic Safety Administration, DOT HS 810 594, 2006.

Knapik, Joseph J., Roberto E. Marin, Tyson L. Grier, and Bruce H. Jones, "A Systematic Review of Post-Deployment Injury-Related Mortality Among Military Personnel Deployed to Conflict Zones," *BMC Public Health*, Vol. 9, No. 231, 2009.

Krahé, Barbara, and Ilka Fenske, "Predicting Aggressive Driving Behavior: The Role of Macho Personality, Age, and Power of Car," *Aggressive Behavior*, Vol. 28, No. 1, 2002, pp. 21–29.

Kraus, J. F., C. Anderson, Paul Zador, A. Williams, S. Arzemanian, W. Li, and M. Salatka, "Motorcycle Licensure, Ownership, and Injury Crash Involvement," *American Journal of Public Health*, Vol. 81, No. 2, 1991, pp. 172–176.

Kwan, I., and J. Mapstone, "Visibility Aids for Pedestrians and Cyclists: A Systematic Review of Randomised Controlled Trials," *Accident Analysis and Prevention*, Vol. 36, No. 3, 2004, pp. 305–312.

Kyrychenko, Sergecy and Anne McCartt, "Florida Weakened Motorcycle Helmet Law: Effects on Death Rates in Motorcycle Crashes," *Traffic Injury Prevention*, Vol. 7, No. 1, 2006, pp. 55–60.

Laberge-Nadeau, C., U. Maag, F. Bellavance, S. D. Lapierre, D. Desjardins, S. Messier, and A. Saidi, "Wireless Telephones and the Risk of Road Crashes," *Accident Analysis and Prevention*, Vol. 35, No. 5, September 2003, pp. 649–660.

Landry, P. R., "SGI's Five Year Occupant Restraint Initiative and Selective Traffic Enforcement Program," *National Leadership Conference on Increasing Safety-Belt Use in the U.S.*, Washington, D.C., 1991, pp. 61–64.

Lapham, Sandra C., Elizabeth Smith, Janet C'De Baca, Iyiin Chang, Betty J. Skipper, George Baum, and William C. Hunt, "Prevalence of Psychiatric Disorders Among Persons Convicted of Driving While Impaired," *Archives of General Psychiatry*, Vol. 58, No. 10, October 1, 2001, pp. 943–949.

Lardelli-Claret, P., J. J. Jimenez-Moleon, J. D. Luna-del-Castillo, M. Garcia-Martin, A. Bueno-Cavanillas, and R. Galvez-Vargas, "Driver Dependent Factors and the Risk of Causing a Collision for Two Wheeled Motor Vehicles," *Injury Prevention*, Vol. 11, No. 4, August 2005, pp. 225–231.

Lawrence, B., W. Max, and T. Miller, *Costs of Injuries Resulting from Motorcycle Crashes: A Literature Review*, Washington, D.C.: National Highway Traffic Safety Administration, U.S. Department of Transportation, DOT HS 809 242, 2002. As of December 20, 2009:
http://www.nhtsa.dot.gov/PEOPLE/INJURY/pedbimot/motorcycle/Motorcycle_HTML/trd.html

Lin, M. R., and J. F. Kraus, "A Review of Risk Factors and Patterns of Motorcycle Injuries," *Accident Analysis and Prevention*, Vol. 41, No. 4, July 2009, pp. 710–722.

Lincoln, Andrew E., Tomoko I. Hooper, Han K. Kang, Samar F. DeBakey, David N. Cowan, and Gary D. Gackstetter, "Motor Vehicle Fatalities Among Gulf War Era Veterans: Characteristics, Mechanisms, and Circumstances," *Traffic Injury Prevention*, Vol. 7, No. 1, 2006, pp. 31–37.

Lucker, G. William, David J. Kruzich, Michael T. Holt, and Jack D. Gold, "The Prevalence of Antisocial Behavior Among U.S. Army DWI Offenders," *Journal of Studies on Alcohol and Drugs*, Vol. 52, No. 4, July 1991, pp. 318–320.

Machin, M. Anthony, and Kim S. Sankey, "Relationships Between Young Drivers' Personality Characteristics, Risk Perceptions, and Driving Behaviour," *Accident Analysis and Prevention*, Vol. 40, No. 2, March 2008, pp. 541–547.

Males, Mike, "The Role of Poverty in California Teenagers' Fatal Traffic Crash Risk," *Californian Journal of Health Promotion*, Vol. 7, No. 1, 2009, pp. 1–13.

Mast, M. S., M. Sieverding, M. Esslen, K. Graber, and L. Jancke, "Masculinity Causes Speeding in Young Men," *Accident Analysis and Prevention*, Vol. 40, No. 2, March 2008, pp. 840–842.

Matthews, G., A. Tsuda, G. Xin, and Y. Ozeki, "Individual Differences in Driver Stress Vulnerability in a Japanese Sample," *Ergonomics*, Vol. 42, 1999, pp. 401–415.

McCartt, Anne T., *Driven to Distraction: Technological Devices and Vehicle Safety: Statement Before the Joint Hearing of the Subcommittee on Commerce, Trade, and Consumer Protection and the Subcommittee on Communications, Technology, and the Internet of the U.S. House of Representatives*, Washington, D.C., November 4, 2009.

McCartt, Anne T., S. A. Ribner, Allan I. Pack, and M. C. Hammer, "The Scope and Nature of the Drowsy Driving Problem in New York State," *Accident Analysis and Prevention*, Vol. 28, No. 4, 1996, pp. 511–517.

McFall, Miles E., Priscilla W. Mackay, and Dennis M. Donovan, "Combat-Related Posttraumatic Stress Disorder and Severity of Substance Abuse in Vietnam Veterans," *Journal of Studies on Alcohol*, Vol. 53, No. 4, 1992, pp. 357–363.

McGwin, G., J. Whatley, J. Metzger, F. Valent, F. Barbone, and L. W. Rue, "The Effect of State Motorcycle Licensing Laws on Motorcycle Driver Mortality Rates," *Journal of Trauma-Injury Infection and Critical Care*, Vol. 56, No. 2, February 2004, pp. 415–419.

Miles, Donna, "Motorcycle, Vehicle Accidents Dominate Off-Duty Summer Fatalities," July 11, 2008. As of January 25, 2009:
http://www.defense.gov/news/newsarticle.aspx?id=50483

Milne, Sarah, Paschal Sheeran, and Sheina Orbell, "Prediction and Intervention in Health-Related Behavior: A Meta-Analytic Review of Protection Motivation Theory," *Journal of Applied Social Psychology*, Vol. 30, No. 1, January 2000, pp. 106–143.

Moore, M., and E. R. Dahlen, "Forgiveness and Consideration of Future Consequences in Aggressive Driving," *Accident Analysis and Prevention*, Vol. 40, No. 5, September 2008, pp. 1661–1666.

Moore, Roland S., and Genevieve M. Ames, "Young Impaired Drivers: The Nature of the Problem and the Strategies Being Used in the Military," paper presented at Young Impaired Drivers: The Nature of the Problem and Possible Solutions, A Workshop, Woods Hole, Mass., June 3–4, 2009.

Morris, C. Craig, *Motorcycle Trends in the United States*, Washington, D.C.: Bureau of Transportation Statistics, U.S. Department of Transportation, SR-014, 2009.

Motor Cycle Council of NSW, Inc., "European Standards for Motorcycle Clothing (PPE)," Colyton, NSW, Australia: Motor Cycle Council of NSW, 2009. As of July 16, 2010:
http://roadsafety.mccofnsw.org.au/a/93.html

Motorcycle Safety Foundation (MSF), "Seasoned Rider Module Fact Sheet," Irvine, Calif.: MSF, 2005.

———, "State Motorcycle Operator Licensing," Irvine, Calif.: MSF, 2008. As of July 16, 2010:
http://www.msf-usa.org/downloads/State_Motorcycle_Operator_Licensing_CSI_2008.pdf

Muller, A., "An Evaluation of the Effectiveness of Motor Cycle Daytime Headlight Laws," *American Journal of Public Health*, Vol. 72, No. 10, 1982, pp. 1136–1141.

Muller, Andreas, "Florida's Motorcycle Helmet Law Repeal and Fatality Rates," *American Journal of Public Health*, Vol. 94, No. 4, 2004, pp. 556–558.

Mullin, Bernadette, Rodney Jackson, John Langley, and Robyn Norton, "Increasing Age and Experience: Are Both Protective Against Motorcycle Injury? A Case-Control Study," *Injury Prevention*, Vol. 6, No. 1, 2000, pp. 32–35.

Mundt, Marlon P., Michael T. French, M. Christopher Roebuck, Linda Baier Manwell, and Kristen Lawton Barry, "Brief Physician Advice for Problem Drinking Among Older Adults: An Economic Analysis of Costs and Benefits," *Journal of Studies on Alcohol*, Vol. 66, 2005.

National Center for Statistics and Analysis (NCSA), *Occupant Protection*, Washington, D.C.: National Highway Traffic Safety Administration, U.S. Department of Transportation, DOT HS 810 621, 2006.

———, *Alcohol-Impaired Driving*, Washington, D.C.: National Highway Traffic Safety Administration, U.S. Department of Transportation, DOT HS 811 155, 2009a.

———, *Seat Belt Use in 2008—Use Rates in the States and Territories*, Washington, D.C.: National Highway Traffic Safety Administration, U.S. Department of Transportation, DOT HS 811 106, April 2009b.

———, *An Examination of Driver Distraction as Recorded in NHTSA Databases*, Washington, D.C.: National Highway Traffic Safety Administration, U.S. Department of Transportation, DOT HS 811 216, September 2009c.

———, *Research Note: Motorcycle Helmet Use in 2009—Overall Results*, Washington, D.C.: U.S. Department of Transportation, National Highway Traffic Safety Administration, DOT HS 811 254, December 2009d.

———, *Early Estimate of Motor Vehicle Traffic Fatalities in 2009*, Washington, D.C.: National Highway Traffic Safety Administration, DOT HS 811 291, March 2010.

National Highway Traffic Safety Administration (NHTSA), *Benefits of Safety Belts and Motorcycle Helmets: Report to Congress—February 1996*, Washington, D.C.: National Highway Traffic Safety Administration, U.S. Department of Transportation, 1996. As of January 25, 2009:
http://www-nrd.nhtsa.dot.gov/Pubs/808-347.pdf

———, *Traffic Safety Facts 2002: Motorcycles*, Washington, D.C.: National Highway Traffic Safety Administration, U.S. Department of Transportation, DOT HS 809 609, 2002. As of January 1, 2010:
http://www-nrd.nhtsa.dot.gov/Pubs/2002mcyfacts.pdf

———, *Motorcycle Safety Program*, Washington, D.C.: National Highway Traffic Safety Administration, U.S. Department of Transportation, DOT HS 809 539, 2003. As of December 22, 2009:
http://204.68.195.151/people/injury/pedbimot/motorcycle/motorcycle03/McycleSafetyProgram.pdf

———, *Traffic Safety Facts, 2003 Data: Motorcycles*, Washington, D.C.: National Highway Traffic Safety Administration, U.S. Department of Transportation, 2004. As of January 1, 2010:
http://www-nrd.nhtsa.dot.gov/Pubs/809764.pdf

———, *Traffic Safety Facts, 2004 Data: Motorcycles*, Washington, D.C.: National Highway Traffic Safety Administration, U.S. Department of Transportation, DOT HS 809 908, 2005a. As of January 1, 2010:
http://www-nrd.nhtsa.dot.gov/Pubs/809908.pdf

———, *Promising Practices in Motorcycle Rider Education and Licensing*, Washington, D.C.: National Highway Traffic Safety Administration, U.S. Department of Transportation, DOT HS 809 922, 2005b. As of January 25, 2009:
http://www.nhtsa.dot.gov/people/injury/pedbimot/motorcycle/MotorcycleRider/index.html

———, *Motorcycle Safety Program*, Washington, D.C.: National Highway Traffic Safety Administration, U.S. Department of Transportation, DOT HS 810 615, 2006a. As of December 15, 2009:
http://www.nhtsa.dot.gov/people/injury/pedbimot/motorcycle/motorcycle03/index.htm

———, *Traffic Safety Facts, 2005 Data: Motorcycles*, Washington, D.C.: National Highway Traffic Safety Administration, U.S. Department of Transportation, DOT HS 810 620, 2006b. As of January 1, 2010:
http://www-nrd.nhtsa.dot.gov/Pubs/810620.pdf

———, *Guidelines for Developing a High-Visibility Enforcement Campaign to Reduce Unsafe Driving Behaviors Among Drivers of Passenger and Commercial Motor Vehicles: A Selective Traffic Enforcement Program (STEP) Based on the Ticketing Aggressive Cars and Trucks (TACT) Pilot Project*, Washington, D.C.: National Highway Traffic Safety Administration, U.S. Department of Transportation, HS810851, 2007a.

———, *Traffic Safety Facts Research Note: Summary of Novelty Helmet Performance Testing*, Washington, D.C.: National Highway Traffic Safety Administration, U.S. Department of Transportation, DOT HS 810 752, 2007b.

———, *Administrative License Revocation*, Washington, D.C.: National Highway Traffic Safety Administration, U.S. Department of Transportation, DOT HS 810 878, January 2008a.

———, *Traffic Safety Facts, 2007 Data: Motorcycles*, Washington, D.C.: National Highway Traffic Safety Administration, U.S. Department of Transportation, DOT HS 810 990, 2008b. As of December 15, 2009:
http://www-nrd.nhtsa.dot.gov/Pubs/810990.pdf

———, *Traffic Safety Facts, Research Note: Motorcycle Helmet Use in 2008—Overall Results*, Washington, D.C.: National Highway Traffic Safety Administration, U.S. Department of Transportation, DOT HS 811 044, 2008c. As of December 15, 2009:
http://www-nrd.nhtsa.dot.gov/Pubs/811044.pdf

———, *Traffic Safety Facts, 2008 Data: Motorcycles*, Washington, D.C.: National Highway Traffic Safety Administration, U.S. Department of Transportation, DOT HS 811 159, 2009a. As of December 15, 2009:
http://www-nrd.nhtsa.dot.gov/Pubs/811159.pdf

———, *Traffic Safety Facts, Research Note: Motorcyclists Injured in Motor Vehicle Traffic Crashes*, Washington, D.C.: National Highway Traffic Safety Administration, U.S. Department of Transportation, DOT HS 811 149, 2009b. As of December 15, 2009:
http://www-nrd.nhtsa.dot.gov/Pubs/811149.pdf

———, *Fatality Analysis Reporting System (FARS) Data Tables*, "National Statistics," Washington, D.C.: National Highway Traffic Safety Administration, U.S. Department of Transportation, 2009c. As of October 2, 2009:
http://www-fars.nhtsa.dot.gov/Main/index.aspx

National Highway Traffic Safety Administration (NHTSA) and Motorcycle Safety Foundation (MSF), *National Agenda for Motorcycle Safety*, Washington, D.C.: National Highway Traffic Safety Administration, U.S. Department of Transportation, November 2000.

———, *National Agenda for Motorcycle Safety (NAMS): Implementation Guide*, Washington, D.C.: National Highway Traffic Safety Administration, U.S. Department of Transportation, DOT HS 810 680, 2006. As of December 15, 2009:
http://www.nhtsa.dot.gov/people/injury/pedbimot/motorcycle/NAMS2006/images/ImplementationGuide.pdf

National Safety Council, *Attributable Risk Estimate Model*, Itasca, Ill.: National Safety Council, 2010.

Naval Safety Center, "Motor Vehicle Tables," Norfolk, Va.: Naval Safety Center, 2010.

Nichols, James L., Neil K. Chaudhary, and Julie Tison, "The Potential for Nighttime Enforcement and Seat Belt Law Upgrades to Impact Alcohol-Related Deaths Among High-Risk Occupants," paper presented at Young Impaired Drivers: The Nature of the Problem and Possible Solutions, A Workshop, Woods Hole, Mass., June 3–4, 2009.

Nilsson, G., *Traffic Safety Dimensions and the Power Model to Describe the Effect of Speed on Safety*, Lund, Sweden: Lund Institute of Technology, Bulletin 221, 2004.

Oltedal, Sigve, and Torbjørn Rundmo, "The Effects of Personality and Gender on Risky Driving Behaviour and Accident Involvement," *Safety Science*, Vol. 44, No. 7, 2006, pp. 621–628.

O'Malley, P. M., and A. C. Wagenaar, "Effects of Minimum Drinking Age Laws on Alcohol Use, Related Behaviors, and Traffic Crash Involvement Among American Youth," *Journal of Studies on Alcohol*, Vol. 52, 1991, pp. 478–491.

Otte, D., G. Schroeder, and M. Richter, "Possibilities for Load Reductions Using Garment Leg Protectors for Motorcyclists—A Technical, Medical and Biomechanical Approach," *Annual Proceedings, Association for the Advancement of Automotive Medicine*, Vol. 46, 2002, pp. 367–385.

Özkan, Türker, and Timo Lajunen, "What Causes the Differences in Driving Between Young Men and Women? The Effects of Gender Roles and Sex on Young Drivers' Driving Behaviour and Self-Assessment of Skills," *Transportation Research Part F: Traffic Psychology and Behaviour*, Vol. 9, No. 4, July 2006, pp. 269–277.

Paine, M., D. Paine, J. Haley, and S. Cockfield, "Daytime Running Lights for Motorcycles," paper presented at 19th International Conference on the Enhanced Safety of Vehicles, Washington, D.C., 2005.

Palmgreen, Philip, Lewis Donohew, Elizabeth Pugzles Lorch, Rick H. Hoyle, and Michael T. Stephenson, "Television Campaigns and Sensation Seeking Targeting of Adolescent Marijuana Use: A Controlled Time Series Approach," in Robert C. Hornik, ed., *Public Health Communication: Evidence for Behavior Change*, Mahwah, N.J.: Lawrence Erlbaum Associates, Inc., 2002, pp. 35–56.

Parker, Dianne, Antony S. R. Manstead, Stephen G. Stradling, James T. Reason, and James S. Baxter, "Intention to Commit Driving Violations: An Application of the Theory of Planned Behavior," *Journal of Applied Psychology*, Vol. 77, No. 1, 1992, pp. 94–101.

Patil, Sujata M., Jean Thatcher Shope, Trivellore E. Raghunathan, and C. Raymond Bingham, "The Role of Personality Characteristics in Young Adult High-Risk Driving," *Traffic Injury Prevention*, Vol. 7, No. 4, 2006, pp. 328-334.

Peck, Raymond C., Michael A. Gebers, Robert B. Voas, and Eduardo Romano, "The Relationship Between Blood Alcohol Concentration (BAC), Age, and Crash Risk," *Journal of Safety Research*, Vol. 39, No. 3, 2008, pp. 311–319.

Philip, Pierre, Jacques Taillard, Patricia Sagaspe, Cedric Valtat, Montserrat Sanchez-Ortuno, Nicholas Moore, Andre Charles, and Bernard Bioulac, "Age, Performance and Sleep Deprivation," *Journal of Sleep Research*, Vol. 13, 2004, pp. 105–110.

Powell, K. E., L. A. Fingerhut, C. M. Branche, and D. M. Perrotta, "Deaths Due to Injury in the Military," *American Journal of Preventive Medicine*, Vol. 18, No. 3, April 2000, pp. 26–32.

Preusser, D. F., J. H. Hedlund, and R. G. Ulmer, *Evaluation of Motorcycle Helmet Law Repeal in Arkansas and Texas*, Washington, D.C.: National Highway Traffic Safety Administration, U.S. Department of Transportation, DOT HS 809 131, 2000. As of January 31, 2010:
http://www.nhtsa.dot.gov/people/injury/pedbimot/motorcycle/EvalofMotor.pdf

Puentes, Robert, and Adie Tomer, *The Road . . . Less Traveled: An Analysis of Vehicle Miles Traveled Trends in the U.S.*, Washington, D.C.: Brookings Institution, 2008.

Radun, Igor, and Jenni E. Radun, "Convicted of Fatigued Driving: Who, Why and How?" *Accident Analysis and Prevention*, Vol. 41, No. 4, July 2009, pp. 869–875.

Radun, Igor, Jenni E. Radun, Heikki Summala, and Mikael Sallinen, "Fatal Road Accidents Among Finnish Military Conscripts: Fatigue-Impaired Driving," *Military Medicine*, Vol. 172, 2007, pp. 1204–1210.

Randolph, Whitney, and K. Viswanath, "Lessons Learned from Public Health Mass Media Campaigns: Marketing Health in a Crowded Media World," *Annual Review of Public Health*, Vol. 25, No. 1, 2004, pp. 419–437.

Reed, Nick, and Ryan Robbins, *The Effect of Text Messaging on Driver Behaviour: A Simulator Study*, London, United Kingdom: RAC Foundation, Published Project Report 367, 2008.

Reeder, A. I., J. C. Alsop, J. D. Langley, and A. C. Wagenaar, "An Evaluation of the General Effect of the New Zealand Graduated Driver Licensing System on Motorcycle Traffic Crash Hospitalisations," *Accident Analysis and Prevention*, Vol. 31, No. 6, November 1999, pp. 651–661.

Reister-Hartsell, M., and R. Gardiner, *Demonstration and Validation of Efficacy of Breath Alcohol Indicators/ Detectors as an Effective Leadership Tool for Reinforcement of "Good Order & Discipline" in Military Units*, Johnstown, Pa.: National Defense Center for Energy and Environment, June 9, 2008.

Retting, Richard A., Robert G. Ulmer, and Allan F. Williams, "Prevalence and Characteristics of Red Light Running Crashes in the United States," *Accident Analysis and Prevention*, Vol. 31, 1999, pp. 687–694.

Roehrs, Timothy, David Beare, Frank Zorick, and Thomas Roth, "Sleepiness and Ethanol Effects on Simulated Driving," *Alcoholism: Clinical and Experimental Research*, Vol. 18, No. 1, 1994, pp. 154–158.

Rosenbloom, Tova, "Sensation Seeking and Risk Taking in Mortality Salience," *Personality and Individual Differences*, Vol. 35, No. 8, 2003, pp. 1809–1819.

Rosencrantz, H., "Rational Policy Goals: Road Safety in Scandinavia," in C. A. Brebbia and V. Dolezel, eds., *Transport XII: Urban Transport and the Environment in the 21st Century*, Southampton, UK: WIT Press, 2006, pp. 151–157.

Ross, David J., "The Prevention of Leg Injuries in Motorcycle Accidents," *Injury*, Vol. 15, No. 2, 1983, pp. 75–77.

Rothman, Alexander J., and Peter Salovey, "Shaping Perceptions to Motivate Healthy Behavior: The Role of Message Framing," *Psychological Bulletin*, Vol. 121, No. 1, 1997, pp. 3–19.

Rumar, Kåre, *Functional Requirements for Daytime Running Lights*, Ann Arbor, Mich.: University of Michigan Transportation Research Institute, UMTRI-2003-11, May 2003. As of December 20, 2009:
http://deepblue.lib.umich.edu/bitstream/2027.42/55184/1/UMTRI-2003-11.pdf

Rutter, D. R., and L. Quine, "Age and Experience in Motorcycling Safety," *Accident Analysis and Prevention*, Vol. 28, No. 1, January 1996, pp. 15–21.

Ryb, Gabriel E., Patricia C. Dischinger, Joseph A. Kufera, and Kathy M. Read, "Risk Perception and Impulsivity: Association with Risky Behaviors and Substance Abuse Disorders," *Accident Analysis and Prevention*, Vol. 38, No. 3, 2006, pp. 567–573.

Sabel, Jennifer C., Lillian S. Bensley, and Juliet van Eenwyk, "Associations Between Adolescent Drinking and Driving Involvement and Self-Reported Risk and Protective Factors in Students in Public Schools in Washington State," *Journal of Studies on Alcohol*, Vol. 65, No. 2, 2004.

Saltz, Robert F., "The Roles of Bars and Restaurants in Preventing Alcohol-Impaired Driving: An Evaluation of Server Intervention," *Evaluation and the Health Professions*, Vol. 10, No. 1, 1987, pp. 5–27.

Salzberg, Philip M., and John M. Moffat, "Ninety Five Percent: An Evaluation of Law, Policy, and Programs to Promote Seat Belt Use in Washington State," *Journal of Safety Research*, Vol. 35, No. 2, 2004, pp. 215–222.

Savolainen, P., and F. Mannering, "Effectiveness of Motorcycle Training and Motorcyclists' Risk-Taking Behavior," *Transportation Research Record*, No. 2031, 2007, pp. 52–58.

Shaffer, Howard J., Sarah E. Nelson, Debi A. LaPlante, Richard A. LaBrie, Mark Albanese, and Gabriel Caro, "The Epidemiology of Psychiatric Disorders Among Repeat DUI Offenders Accepting a Treatment-Sentencing Option," *Journal of Consulting and Clinical Psychology*, Vol. 75, No. 5, 2007, pp. 795–804.

Shankar, Umesh, *Fatal Single Vehicle Motorcycle Crashes*, Washington, D.C.: National Highway Traffic Safety Administration, Mathematical Analysis Division, U.S. Department of Transportation, DOT HS 809 360, 2001. As of January 20, 2010:
http://www-nrd.nhtsa.dot.gov/Pubs/809-360.pdf

Shankar, Umesh G., *Research Note: Alcohol Involvement in Fatal Motorcycle Crashes*, Washington, D.C.: National Highway Traffic Safety Administration, U.S. Department of Transportation, DOT HS 809 576, 2003.

Shankar, Umesh, and Cherian Varghese, *Recent Trends in Fatal Motorcycle Crashes: An Update*, Washington, D.C.: National Highway Traffic Safety Administration, U.S. Department of Transportation, DOT HS 810 606, 2006. As of January 8, 2010:
http://www-nrd.nhtsa.dot.gov/Pubs/810606.pdf

Shankar, Umesh, and Keith Wardell, *Geo-Demographic Analysis of Fatal Motorcycle Crashes*, Washington, D.C.: National Highway Traffic Safety Administration, U.S. Department of Transportation, DOT HS 809 197, 2001. As of December 16, 2009:
http://www-nrd.nhtsa.dot.gov/Pubs/809197.pdf

Shope, Jean T., and C. Raymond Bingham, "Teen Driving—Motor-Vehicle Crashes and Factors That Contribute," *American Journal of Preventive Medicine*, Vol. 35, No. 3, September 2008, pp. S261–S271.

Shults, Ruth A., Randy W. Elder, David A. Sleet, James L. Nichols, Mary O. Alao, Vilma G. Carande-Kulis, Stephanie Zaza, Daniel M. Sosin, Robert S. Thompson, and the Task Force on Community Preventive Services, "Task Force on Community Preventive Services: Reviews of Evidence Regarding Interventions to Reduce Alcohol-Impaired Driving," *American Journal of Preventive Medicine*, Vol. 21, No. 4S, 2001, pp. 66–88.

Sibley, Chris G., and Niki Harré, "The Impact of Different Styles of Traffic Safety Advertisement on Young Drivers' Explicit and Implicit Self-Enhancement Biases," *Transportation Research Part F: Traffic Psychology and Behaviour*, Vol. 12, No. 2, March 2009, pp. 159–167.

Sivak, Michael, Juha Luoma, Michael J. Flannagan, C. Raymond Bingham, David W. Eby, and Jean T. Shope, *Traffic Safety in the US: Re-Examining Major Opportunities*, Ann Arbor, Mich.: University of Michigan Transportation Research Institute, UMTRI-2006-26, 2006.

Smith, Simon S., Mark S. Horswill, Brooke Chambers, and Mark Wetton, "Hazard Perception in Novice and Experienced Drivers: The Effects of Sleepiness," *Accident Analysis and Prevention*, Vol. 41, No. 4, July 2009, pp. 729–733.

Soderstrom, Carl A., Patricia C. Dischinger, Timothy J. Kerns, and Anna L. Trifillis, "Marijuana and Other Drug Use Among Automobile and Motorcycle Drivers Treated at a Trauma Center," *Accident Analysis and Prevention*, Vol. 27, No. 1, 1995, pp. 131–135.

Soderstrom, Carl A., Anna L. Trifillis, Belavadi S. Shankar, William E. Clark, and R. Adams Cowley, "Marijuana and Alcohol Use Among 1023 Trauma Patients: A Prospective Study," *Archives of Surgery*, Vol. 123, No. 6, June 1, 1988, pp. 733–737.

Solomon, M. G., N. K. Chaudhary, and L. A. Cosgrove, *May 2003 Click It or Ticket Safety Belt Mobilization Campaign*, Washington, D.C.: U.S. Department of Transportation, 2003.

Solomon, M. G., Stephanie H. Gilbert, James L. Nichols, Robert H. B. Chaffe, and Neil K. Chaudhary, *Evaluation of the May 2005 Click It or Ticket Mobilization to Increase Seat Belt Use*, Washington, D.C.: National Highway Traffic Safety Administration, U.S. Department of Transportation, 2007.

Sosin, Daniel M., Jeffrey J. Sacks, and Patricia Holmgreen, "Head Injury—Associated Deaths from Motorcycle Crashes: Relationship to Helmet-Use Laws," *Journal of the American Medical Association*, Vol. 264, No. 18, November 14, 1990, pp. 2395–2399.

Stahre, Mandy A., Robert D. Brewer, Vincent P. Fonseca, and Timothy S. Naimi, "Binge Drinking Among U.S. Active-Duty Military Personnel," *American Journal of Preventive Medicine*, Vol. 36, No. 3, 2009, pp. 208–217.

Stead, Martine, Stephen Tagg, Anne Marie MacKintosh, and Douglas Eadie, "Development and Evaluation of a Mass Media Theory of Planned Behaviour Intervention to Reduce Speeding," *Health Education Research*, Vol. 20, No. 1, February 2005, pp. 36–50.

Steinberg, Laurence, "Should the Science of Adolescent Brain Development Inform Public Policy?" *American Psychologist*, Vol. 64, No. 8, November 2009, pp. 739–750.

Stern, Erica, and Todd Rockwood, *Post-Deployment Driving Problems: Survey of Scope and Timeline for Post-Deployment Soldiers With and Without Mild Traumatic Brain Injury*, Twin Cities: University of Minnesota, forthcoming.

Strohl, Kingman P., Jesse Blatt, Forrest Council, Kate Georges, James Kiley, Roger Kurrus, Anne T. McCartt, Sharon L. Merritt, Allan I. Pack, Susan Rogus, Thomas Roth, Jane Stutts, Pat Waller, and David Willis, *Drowsy Driving and Automobile Crashes*, Washington, D.C.: NCSDR/NHTSA Expert Panel on Driver Fatigue and Sleepiness, National Highway Traffic Safety Administration, U.S. Department of Transportation, 1998. As of June 25, 2010:
http://www.nhtsa.dot.gov/people/injury/drowsy_driving1/Drowsy.html#NCSDR/NHTSA

Stuster, Jack W., *The Detection of DWI Motorcyclists*, Washington, D.C.: National Highway Traffic Safety Administration, U.S. Department of Transportation, DOT HS 807 839, 1993.

———, *The Detection of DWI at BACs Below 0.10 (Final Report)*, Washington, D.C.: National Highway Traffic Safety Administration, U.S. Department of Transportation, September 12, 1997.

———, *Aggressive Driving Enforcement: Evaluations of Two Demonstration Programs*, Washington, D.C.: National Highway Traffic Safety Administration, DOT HS 809 707, 2004. As of July 24, 2010:
http://ntl.bts.gov/lib/24000/24600/24654/AggresDrvngEnforce-5.0.pdf

Stuster, Jack W., and Zail Coffman, *Synthesis of Safety Research Related to Speed and Speed Limits*, Washington, D.C.: Federal Highway Administration, FHWA RD 98 154, 1998. As of June 25, 2010:
http://www.tfhrc.gov/safety/speed/speed.htm

Stutts, Jane, John Feaganes, Donald Reinfurt, Eric Rodgman, Charles Hamlett, Kenneth Gish, and Loren Staplin, "Driver's Exposure to Distractions in Their Natural Driving Environment," *Accident Analysis and Prevention*, Vol. 37, No. 6, 2005a, pp. 1093–1101.

Stutts, Jane, Ronald R. Knipling, Ronald Pfefer, Timothy R. Neuman, Kevin L. Slack, and Kelly K. Hardy, *Volume 14: A Guide for Reducing Crashes Involving Drowsy and Distracted Drivers*, Washington, D.C.: Transportation Research Board, NCHRP Report 500, 2005b.

Subramanian, Rajesh, *Motor Vehicle Traffic Crashes as a Leading Cause of Death in the U.S., 2002—A Demographic Perspective*, Washington, D.C.: National Highway Traffic Safety Administration, U.S. Department of Transportation, DOT HS 809 843, 2005.

Sugano, Dean, *Cell Phone Use and Motor Vehicle Collisions: A Review of the Studies*, Honolulu: Legislative Reference Bureau, Report No. 4, November 2005.

Sun, Stephen W., David M. Kahn, Kenneth G. Swan, "Lowering the Legal Blood Alcohol Level for Motorcyclists," *Accident Analysis and Prevention*, Vol. 30, 1998, pp. 133–136.

Svenson, O., "Are We Less Risky and More Skillful Than Our Fellow Drivers?" *Acta Psychologica*, Vol. 47, No. 2, 1981, pp. 143–148.

Tanielian, Terri, and Lisa H. Jaycox, eds., *Invisible Wounds of War: Psychological and Cognitive Injuries, Their Consequences, and Services to Assist Recovery*, Santa Monica, Calif.: RAND Corporation, MG-720-CCF, 2008. As of June 25, 2010:
http://www.rand.org/pubs/monographs/MG720/

Thuen, F., "Injury-Related Behaviours and Sensation Seeking: An Empirical Study of a Group of 14 Year Old Norwegian School Children," *Health Education Research*, Vol. 9, No. 4, December 1, 1994, pp. 465–472.

Transportation Research Board, "Young Impaired Drivers: The Nature of the Problem and Possible Solutions," Washington, D.C.: Transportation Research Board, Transportation Research Circular No. E-C132, 2009.

Ulleberg, Pål, "Personality Subtypes of Young Drivers. Relationship to Risk-Taking Preferences, Accident Involvement, and Response to a Traffic Safety Campaign," *Transportation Research Part F: Traffic Psychology and Behaviour*, Vol. 4, No. 4, 2001, pp. 279–297.

Ulleberg, Pål, and Torbjørn Rundmo, "Personality, Attitudes and Risk Perception as Predictors of Risky Driving Behaviour Among Young Drivers," *Safety Science*, Vol. 41, No. 5, June 2003, pp. 427–443.

Ulmer, Robert G., and Veronika Shabanova Northrup, *Evaluation of the Repeal of the All-Rider Motorcycle Helmet Law in Florida*, Washington, D.C.: National Highway Traffic Safety Administration, U.S. Department of Transportation, DOT HS 809 849, 2005. As of January 29, 2010:
http://www.nhtsa.dot.gov/people/injury/pedbimot/motorcycle/flamcreport/pages/index.htm

U.S. Army Safety Management Information System (ASMIS), *U.S. Army Accident Information, Army Historical Statistical Report*, "Ground Accidents," various years. As of June 25, 2010:
https://rmis.army.mil/stats/prc_army_stats_history/

Vanderbilt, Tom, *Traffic: Why We Drive the Way We Do*, New York/Toronto: Alfred A. Knopf, 2008.

Villaveces, Andrés, Peter Cummings, Thomas D. Koepsell, Frederick P. Rivara, Thomas Lumley, and John Moffat, "Association of Alcohol-Related Laws with Deaths Due to Motor Vehicle and Motorcycle Crashes in the United States, 1980–1997," *American Journal of Epidemiology*, Vol. 157, 2003, pp. 131–140.

Vivoda, Jonathon M., David W. Eby, Renée M. St. Louis, and Lidia P. Kostyniuk, "A Direct Observation Study of Nighttime Safety Belt Use in Indiana," *Journal of Safety Research*, Vol. 38, No. 4, 2007, pp. 423–429.

Voas, R. B., A. S. Tippetts, and J. C. Fell, "Assessing the Effectiveness of Minimum Legal Drinking Age and Zero Tolerance Laws in the United States," *Accident Analysis and Prevention*, Vol. 35, 2003, pp. 579–587.

Wagenaar, Alexander C., and Mildred M. Maldonado-Molina, "Effects of Drivers' License Suspension Policies on Alcohol-Related Crash Involvement: Long-Term Follow-up in Forty-Six States," *Alcoholism Clinical and Experimental Research*, Vol. 31, No. 8, 2007.

Wagenaar, A. C., and A. L. Tobler, "Alcohol Sales and Service to Underage Youth and Intoxicated Patrons: Effects of Responsible Beverage Service Training and Enforcement Interventions," *Traffic Safety and Alcohol Regulations: A Symposium*, Transportation Research Circular No. E-C123, 2007.

Walton, D., and J. Bathurst, "An Exploration of the Perceptions of the Average Driver's Speed Compared with Perceived Driver Safety and Driving Skill," *Accident Analysis and Prevention*, Vol. 30, No. 6, 1998, pp. 821–830.

Wang, Jing-Shiarn, *The Effectiveness of Daytime Running Lights for Passenger Vehicles*, Washington, D.C.: National Highway Traffic Safety Administration, U.S. Department of Transportation, DOT HS 811 029, 2008.

Wang, J. S., R. R. Knipling, and M. J. Goodman, "The Role of Driver Inattention in Crashes; New Statistics from the 1995 Crashworthiness Data System," in *Association for the Advancement of Automotive Medicine—40th Annual Proceedings*, Des Plaines, Ill.: Association for the Advancement of Automotive Medicine, 1996, pp. 377–392.

Wells, J. K., D. K. Preusser, and A. F. Williams, "Enforcing Alcohol-Impaired Driving and Seat Belt Use Laws, Binghamton, NY," *Journal of Safety Research*, Vol. 23, 1992.

Wells, Susan, Bernadette Mullin, Robyn Norton, John Langley, Jennie Connor, Roy Lay-Yee, and Rod Jackson, "Motorcycle Rider Conspicuity and Crash Related Injury: Case-Control Study," *British Medical Journal*, Vol. 328, No. 7444, April 2004, pp. 857–860A.

Wells-Parker, Elisabeth, Robert Bangert-Drowns, Robert McMillen, and Marsha Williams, "Final Results from a Meta-Analysis of Remedial Interventions with Drink/Drive Offenders," *Addiction*, Vol. 9, No. 7, 1995, pp. 907–926.

White, Joseph B., "New Reports Shows Significant Drop in Auto Fatalities," *Wall Street Journal*, February 9, 2009.

Williams, A. F., in R. Jessor, ed., *New Perspectives on Adolescent Risk Behaviour*, Cambridge, Mass.: Cambridge University Press, 1998, pp. 221–237.

Williams, Allan F., Michael A. Peat, Dennis J. Crouch, Joann K. Wells, Bryan S. Finkle, "Drugs in Fatally Injured Young Male Drivers," *Public Health Report*, Vol. 100, No. 1, 1985, pp. 19–25.

Williams, Jeffrey O., Nicole S. Bell, and Paul J. Amoroso, "Drinking and Other Risk Taking Behaviors of Enlisted Male Soldiers in the U.S. Army," *Work: A Journal of Prevention, Assessment and Rehabilitation*, Vol. 18, No. 2, 2002, pp. 141–150.

Wilson, Abigail L. Garvey, Jeffrey L. Lange, John F. Brundage, and Robert A. Frommelt, "Behavioral, Demographic, and Prior Morbidity Risk Factors for Accidental Death Among Men: A Case-Control Study of Soldiers," *Preventive Medicine*, Vol. 36, No. 1, January 2003, pp. 124–130.

Wilson, R. J., "The Relationship of Seat Belt Non-Use to Personality, Lifestyle and Driving Record," *Health Education Research*, Vol. 5, No. 2, June 1, 1990, pp. 175–185.

Witte, Kim, and Mike Allen, "A Meta-Analysis of Fear Appeals: Implications for Effective Public Health Campaigns," *Health Education Behavior*, Vol. 27, No. 5, October 1, 2000, pp. 591–615.

Yanovitzky, Itzhak, and Courtney Bennett, "Media Attention, Institutional Response, and Health Behavior Change—The Case of Drunk Driving, 1978–1996," *Communication Research*, Vol. 26, No. 4, August 1999, pp. 429–453.

Young, Kristie L., and Michael Regan, "Driver Distraction: A Review of the Literature," in I. J. Faulks, M. Regan, M. Stevenson, J. Brown, A. Porter, and J. D. Irwin, eds., *Distracted Driving*, Sydney, NSW, Australia: Australasian College of Road Safety, 2007, pp. 379–405.

Yu, Jiang and William R. Williford, "Problem Drinking and High-Risk Driving: An Analysis of Official and Self-Reported Drinking-Driving in New York State," *Addiction*, Vol. 88, No. 2, 1993, pp. 219–228.

Zador, Paul L., "Motorcycle Headlight-Use Laws and Fatal Crashes in the U.S.," *American Journal of Public Health*, Vol. 75, No. 5, 1985, pp. 543–546.

Zador, Paul L., S. A. Krawchuk, and R. B. Voas, *Relative Risk of Fatal and Crash Involvement by BAC, Age and Gender*, Washington, D.C.: National Highway Traffic Safety Administration, U.S. Department of Transportation, DOT HS 809 050, April 2000.

Zambon, F., and M. Hasselberg, "Socioeconomic Differences and Motorcycle Injuries: Age at Risk and Injury Severity Among Young Drivers—A Swedish Nationwide Cohort Study," *Accident Analysis and Prevention*, Vol. 38, No. 6, November 2006, pp. 1183–1189.

Zuckerman, M., *Behavioral Expressions and Biosocial Bases of Sensation Seeking*, New York: Cambridge University Press, 1994.

———, *Sensation Seeking and Risky Behavior*, Washington, D.C.: American Psychological Association, 2007.

Zuckerman, Marvin, and D. Michael Kuhlman, "Personality and Risk-Taking: Common Bisocial Factors," *Journal of Personality*, Vol. 68, No. 6, December 2000, pp. 999–1029.